Classical Control Using H^∞ Methods

An Introduction to Design

Classical Control Using H^∞ Methods

An Introduction to Design

J. William
Helton
University of California
San Diego, California

Orlando
Merino
University of Rhode Island
Kingston, Rhode Island

siam.

Society for Industrial and Applied Mathematics

Philadelphia

Copyright ©1998 by the Society for Industrial and Applied Mathematics.

10 9 8 7 6 5 4 3 2 1

All rights reserved. Printed in the United States of America. No part of this book may be reproduced, stored, or transmitted in any manner without the written permission of the publisher. For information, write to the Society for Industrial and Applied Mathematics, 3600 University City Science Center, Philadelphia, PA 19104-2688.

Typeset by T_EXniques, Inc., Boston, MA and the Society for Industrial and Applied Mathematics. Printed by Victor Graphics, Inc., Baltimore, MD.

Library of Congress Catalog Card Number: 98-86611

siam is a registered trademark.

Contents

Preface ix

I Short Design Course 1

1 A Method for Solving System Design Problems 3
- 1.1 Rational functions 3
- 1.2 The closed-loop system $\mathcal{S}$ 4
- 1.3 Designable transfer function 6
- 1.4 A system design problem 6
- 1.5 The method 7
- 1.6 Exercises 9

2 Internal Stability 11
- 2.1 Control and stability 11
- 2.2 Interpolation 13
- 2.3 Systems with a stable plant 15
- 2.4 Exercise 16

3 Frequency Domain Performance Requirements 17
- 3.1 Introduction 17
  - 3.1.1 The closed-loop system $\mathcal{S}$ 17
  - 3.1.2 Frequency domain performance requirements 18
  - 3.1.3 Disk inequalities 18
- 3.2 Measures of performance 19
  - 3.2.1 Gain-phase margin 19
  - 3.2.2 Tracking error 21
  - 3.2.3 Bandwidth 22
  - 3.2.4 Closed-loop roll-off 23
  - 3.2.5 Fundamental trade-offs 24
  - 3.2.6 Choosing sets of performance requirements 24
- 3.3 Piecing together disk inequalities 25
- 3.4 More performance measures 27
  - 3.4.1 Peak magnitude 27

		3.4.2	Compensator bound	27
		3.4.3	Plant bound	28
		3.4.4	Disturbance rejection	29
		3.4.5	More on tracking error	30
		3.4.6	Tracking and type n plants	31
	3.5	A fully constrained problem		32

4 Optimization   35
    4.1 Review of concepts . . . 35
    4.2 Generating a performance function . . . 36
    4.3 Finding T with best performance . . . 38
       4.3.1 Example . . . 39
    4.4 Acceptable performance functions . . . 40
    4.5 Performance not of the circular type . . . 43
    4.6 Optimization . . . 44
       4.6.1 The optimization problem $OPT_\mathcal{I}$ . . . 44
    4.7 Internal stability and optimization . . . 45
       4.7.1 The optimization problem OPT . . . 45
       4.7.2 OPT with circular Γ . . . 45
    4.8 Exercises . . . 46

5 A Design Example with OPTDesign   47
    5.1 Introduction . . . 47
    5.2 The problem . . . 47
    5.3 Optimization with OPTDesign . . . 50
    5.4 Producing a rational compensator . . . 52
    5.5 How good is the answer? . . . 54
       5.5.1 More on plots and functions . . . 56
    5.6 Optimality diagnostics . . . 58
    5.7 Specifying compensator roll-off . . . 58
    5.8 Reducing the numerical error . . . 59
    5.9 Rational Fits . . . 60
    5.10 Exercises . . . 61

II More on Design   63

6 Examples   65
    6.1 Numerical practicalities . . . 65
       6.1.1 Sampling functions on the $j\omega$ axis . . . 66
       6.1.2 Discontinuous functions . . . 67
       6.1.3 Vanishing radius function . . . 68
       6.1.4 Performance function incorrectly defined . . . 69
    6.2 Design example 1 . . . 69
       6.2.1 Electro-mechanical and electrical models . . . 69
       6.2.2 Mathematical model . . . 71

		6.2.3 Statement of the problem	73
		6.2.4 Reformulation of requirements	73
		6.2.5 Optimization	74
		6.2.6 Second modification of the envelope and optimization	78
	6.3	Time domain performance requirements	81
		6.3.1 Two common time domain requirements	83
		6.3.2 A naive method	83
		6.3.3 A refinement of the naive method	85
	6.4	Design example 2	85
		6.4.1 Statement of the problem	86
		6.4.2 Translation of time domain requirements	86
	6.5	Performance for competing constraints	93
		6.5.1 Rounding corners of performance functions	94
		6.5.2 Constrained optimization with a barrier method	97
7	**Internal Stability II**		**101**
	7.1	Calculating interpolants	101
		7.1.1 Calculating one interpolant	102
		7.1.2 Parameterization of all interpolants	103
		7.1.3 Interpolation with a relative degree condition	104
	7.2	Plants with simple RHP zeros and poles	105
	7.3	Parameterization: The general case	107
		7.3.1 Higher-order interpolation	107
		7.3.2 Plants with high-multiplicity RHP zeros and poles	109
	7.4	Exercises	110

III Appendices   113

A History and Perspective   115

B Downloading OPTDesign and *Anopt*   119

C Computer Code for Example in Chapter 6   121
 C.1 Computer code for design example 1 121
 C.2 Computer code for design example 2 124

D *Anopt* Notebook   127
 D.1 Foreword 127
 D.2 Optimizing in the sup norm: The problem *OPT* 128
 D.3 Example 1: First run 128
 D.4 Example 2: Vector-valued analytic functions 131
 D.5 Example 3: Specification of more input 132
 D.6 Quick reference for *Anopt* 135

E NewtonInterpolant Notebook — 137
- E.1 First calculation of an interpolant — 138
- E.2 Specifying the pole location — 138
- E.3 Specifying of the relative degree — 139
- E.4 Complex numbers as data — 139
- E.5 Higher-order interpolation — 140

F NewtonFit Notebook — 143
- F.1 First example — 144
- F.2 Template for many runs — 147
- F.3 Using a weight — 147
- F.4 Stable zeros and poles — 148

G OPTDesign Plots, Data, and Functions — 151
- G.1 Functions and grids — 151
- G.2 Plots — 153
- G.3 Rational approximation and model reduction — 156

References — 159

Index — 169

Preface

Purpose of this book

One of the main accomplishments of control in the 1980s was the development of H^∞ engineering. It constitutes a systematic methodology for frequency domain design which optimizes worst-case performance specifications. This book teaches control system design using H^∞ techniques and is *aimed at students of control who want an H^∞ supplement to their control course.*

We believe that the approach presented here has three virtues. First, it is conceptually simple. Secondly, it corresponds directly to what is often called "classical control", which is good simply because of the widespread appeal of classical frequency domain methods. Finally, classical control has often been presented as trial and error applied to specific cases; this book lays out a much more precise approach. This has the tremendous advantage of converting an engineering problem to one which can be put directly into a mathematical optimization package. While some trial and error may be needed, our approach greatly reduces it.

A student who learns Part I should have a good feel for how engineering specs are encoded as precise mathematical constraints.

Software

This book also can serve to introduce the control software package OPTDesign, which runs under Mathematica. The book is independent of the software; however, the companionship of the software provides a richer experience. With OPTDesign the reader can easily reproduce the calculations done here in the solved examples, and try variations on them.

Recommended background

The book requires knowing basics of engineering like what a frequency response function is. For example, the book can be well learned by a student who has had a one semester control course, maybe less. It could serve as a supplement to many popular control texts such as [FPE86], [Do81] and [O90].

Optimal design

We shall always deal with a situation where part of a linear system is given (called the plant in control theory). We want to find the additional part f (the designable part) so that the whole system meets certain requirements.

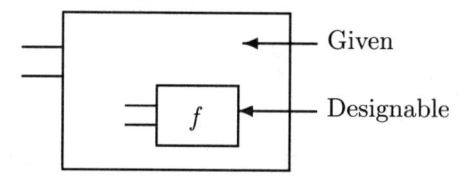

Fig. P.1. *The given and designable parts of the system.*

A high-level description of the design method described in this book is:

I. List requirements carefully and find a mathematical representation for them.

II. Obtain a performance function Γ from the given part of the system and the requirements listed in I, in terms of the designable variable f. The performance function is a function Γ of frequency ω and f. In other words, the performance of the closed loop system at frequency ω is $\Gamma(j\omega, f(j\omega))$. The function Γ is actually a cost function; the smaller its value, the better the design.

III. Optimize the performance to obtain an acceptable system, or conclude that no such system exists with the given requirements.

This book presents a list of requirements that are sound from the physical point of view yet simple mathematically. The list of requirements discussed here is by no means complete: as more is understood about systems, more requirements can be added to those in this book. It is our contention that the framework is flexible enough to accommodate them. Several examples illustrate the practical solution of design problems using steps I–III.

H^∞-Control optimizes the performance over all frequencies, not just averages (mean square) performance over frequencies. This approach is true to the physical problem where the older mean square optimization approaches distort engineering specifications in order to produce a mathematically easy problem. Fortunately in the last 15 years the mathematics of H^∞ has become powerful enough to solve engineering problems.

SISO and MIMO

This book emphasizes single-input–single-output (SISO) systems. However, the methods used here can be generalized readily to multiple-input–multiple-output systems (MIMO). MIMO design problems are not supported by the current

PREFACE

version of the software package OPTDesign. However, the optimization routine Anopt used by OPTDesign has capability of solving many optimization problems of a kind that arise in MIMO design problems.

The long version of the book

This manuscript is exerpted from a longer book which contains five parts plus appendices. The later parts of the book treat the theory of H^∞ control and should be valuable to theoreticians in control and to research mathematicians. Even practicioners might benefit from a light reading of this theory. Since the long book is highly modular, the theory parts can be read independently of the design parts and vice-versa.

Other references

Advanced books that present H^∞ theory but with a different approach include [F87], [GL95], [BGR90], [FF91], [GM90], [DFT92], [H87], [Dy89], [Kim97], [ZDG96]. All of these books are based on the same type of mathematics, namely, on extensions of something called Nevanlinna–Pick, Nehari and commutant lifting theories. Another approach is [BB91]. There exist commercial software packages which do H^∞ control design, such as toolboxes available from The MathWorks Inc. (Robust Control toolbox, LMI Control toolbox, μ-Analysis and Synthesis toolbox, QFT toolbox,) Delight (Prof. André Tits, Univ. of Maryland,) Qdes (Prof. S. Boyd, Stanford University.) The mathematical core of our book presented in the later parts and our software is fundamentally different from both approaches.

Thanks

Thanks to Trent Walker, David Ring, Robert O'Barr, David Schwartz, Neola Crimmins, Joy Kirsch, Rosetta Fuller, Mike Swafford, and John Flattum. John Boyd, Dan Cunningham, Frank Hauser, and Alan Schneider made valuable comments on the exposition. Eric Rowell read a draft of the book and caught many errors.

The index is due to Jeremy Martin. Julia Myers read many drafts of the book and found many typographical and grammatical errors, and cases of unclear exposition. We thank Julia for her wonderful work. We are especially grateful to Prof. Fred Bailey of the University of Minnesota, his student Brett Larson, and to Prof. A. T. Shenton and research assistant Zia Shafiei of the University of Liverpool. Finally, we would like to thank the Air Force Office of Scientific Research, the National Science Foundation and the Ford Motor Co. for partially supporting the writing of this book through grants.

Code development

OPTDesign and its companion Anopt are based on algorithms due to Helton, Merino and Walker, and written in Mathematica by Orlando Merino, Julia

Myers and Trent Walker. Also Daniel Lam, Robert O'Barr, Mark Paskowitz, and Mike Swafford contributed. Earlier versions of Anopt (Fortran, 1989) were designed and written by Jim Bence, J. William Helton, Julia Myers, Orlando Merino, Robert O'Barr and David Ring. An even earlier Fortran package is Approxih by David Schwartz and J. W. Helton (1985).

Downloading OPTDesign and Anopt

The software packages OPTDesign and Anopt can be obtained at the Anopt web site *anopt.math.ucsd.edu*, or by anonymous ftp as described in Appendix B.

J. William Helton   *Orlando Merino*

Part I
Short Design Course

Chapter 1

A Method for Solving System Design Problems

This chapter outlines our approach to solving design problems. Section 1.1 gives basic facts about rational functions, which are the single most important type of functions in practical system design. The basic system and functions considered in this book are introduced in section 1.2 and section 1.3. A control design problem is presented in section 1.4. A description of the method used in this book to solve control problems is presented in section 1.5.

1.1 Rational functions

If $N(s)$ and $D(s)$ are polynomials, then the rational function $F(s) = \frac{N(s)}{D(s)}$ has *relative degree* given by

$$d(F(s)) := \text{degree}(D(s)) - \text{degree}(N(s))$$

The function F is called *proper* if $d(F) \geq 0$ and *strictly proper* if $d(F) > 0$.

A rational function can be thought of as a function on the imaginary axis $F(j\omega)$ or as a function on the complex plane, where the notation $F(s)$ is used. A function F is said to be *bounded* (on the imaginary axis) if there exists a constant $M > 0$ such that

$$|F(j\omega)| \leq M, \quad \text{for all } \omega \in \mathbb{R}.$$

The smallest such M denoted $||F||_\infty$, namely,

$$||F||_\infty := \sup_{\omega \in \mathbb{R}} |F(j\omega)|$$

is called the *supremum* of $|F(j\omega)|$ (see Fig. 1.1).

A rational function F is bounded if and only if it is proper and has no poles on the imaginary axis; the proof of this is left as an exercise. The value $||F||_\infty$ is not always attained, as the example below illustrates.

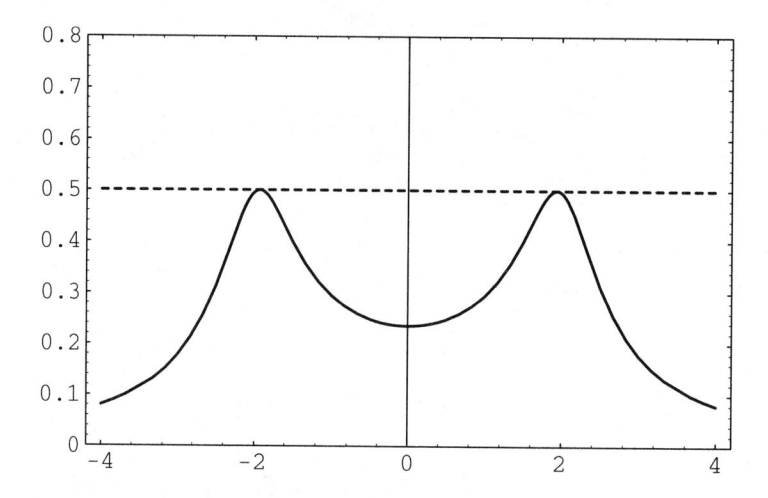

Fig. 1.1. Plot of $|F(j\omega)|$ versus frequency ω, where $F(s) = 1/((s-0.5)^2 + 4)$. The dotted line marks $\|F\|_\infty$.

Example. The function $F(s) = s^2/(s^2 - 10)$ is bounded, since $|F(j\omega)| \leq 1$ for all $\omega \in \mathbb{R}$. However, for this example there is no single finite frequency ω_0 at which $|F(j\omega_0)| = 1$. In other words, the supremum is not attained.

Functions $F(s)$ that satisfy the equality

$$F(s) = \overline{F(\bar{s})}, \tag{1.1}$$

for all s, are sometimes referred to as *real* (on the real axis) and are very important in engineering. They arise as the Laplace transform of real-valued functions of time. On the imaginary axis equality (1.1) becomes

$$F(j\omega) = \overline{F(-j\omega)}. \tag{1.2}$$

All functions that appear in this book have this property. One can show that a rational function is real if and only if the coefficients are real numbers.

A rational function is *stable* if every pole of the function has negative real part and the relative degree is not positive.

The usual convention is to denote the set of all rational functions that are stable and real by $\mathcal{RH}^\infty$. Note that functions in $\mathcal{RH}^\infty$ can be described as proper rational functions with real coefficients and with no poles in the closed right half-plane RHP.[1]

1.2 The closed-loop system $\mathcal{S}$

The basic closed-loop system we consider is a linear, time-invariant, finite-dimensional system in the frequency domain description. It is depicted in Fig.

---

[1] Some of the concepts introduced here apply to nonrational functions as well. For example, a continuous function $F(j\omega)$ may be bounded or may satisfy (1.2).

1.2. THE CLOSED-LOOP SYSTEM $\mathcal{S}$

1.2.

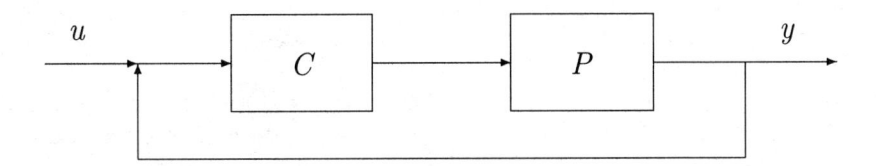

Fig. 1.2. *The closed-loop system $\mathcal{S}$*

The functions P and C are proper, real, rational functions of s. They are called the *plant* and the *compensator*, respectively. Figure 1.2 is called the closed-loop system and denoted by $\mathcal{S}$. We assume that the plant in Fig. 1.2 is given and cannot be modified, and that many choices of the compensator are possible.

Besides P and C, there are other functions commonly associated with the closed-loop system $\mathcal{S}$, which we now derive. Begin with Fig. 1.2 to obtain the equations

$$\begin{aligned} y(s) &= P(s)C(s)e(s) \\ e(s) &= u(s) - y(s). \end{aligned} \tag{1.3}$$

Combine the two equations in (1.3) to obtain

$$y(s) = P(s)C(s)u(s) - P(s)C(s)y(s).$$

Solve for $y(s)$ and obtain

$$y(s) = (1 + P(s)C(s))^{-1} P(s)C(s) u(s). \tag{1.4}$$

Equation (1.4) contains the closed-loop transfer function of $\mathcal{S}$, defined as

$$T := \frac{PC}{1 + PC}. \tag{1.5}$$

Note that the variable s has been suppressed in (1.5) to simplify notation. Other key functions are

- The sensitivity function $S := (1 + PC)^{-1}$;

- The tracking error for input u, Su;

- The closed-loop compensator $Q := CS$;

- The closed-loop plant $C(1 + PC)^{-1}$;

- The open-loop transfer function $L := PC$.

1.3 Designable transfer function

Roughly speaking, to design a system one must choose all of the system's parts that have not been given in advance. This is done according to some criteria or requirements.

The most obvious way to select all the parts of system $\mathcal{S}$ is to specify C. Doing this determines the closed-loop system $\mathcal{S}$ completely, in the sense that all its functions T, L, S, Q, etc. can be written in terms of the given part P and our selection for C.

In this book we will use another way to select a closed-loop system $\mathcal{S}$. Think of T as a variable.[2] If T is specified, then one can calculate C by solving in (1.5):

$$C = \frac{T}{(1-T)P}. \qquad (1.6)$$

Now we see that it is possible to write C and every transfer function associated to $\mathcal{S}$ in terms of T and the given part P. That is, to determine the closed-loop system $\mathcal{S}$ one has only to pick a particular value for T. Under these circumstances we refer to T as a *designable transfer function*

The designable transfer function cannot take an arbitrary form. There are many properties that the physics or mathematics of the system dictate. There is a set $\mathcal{I}$ of admissible functions to which the designable T must belong. For example, $\mathcal{I}$ contains functions that are continuous on the imaginary axis and uniformly bounded there. Clearly, the set $\mathcal{I}$ must be defined before the design process begins. In this book the set $\mathcal{I}$ is a subset of $\mathcal{RH}^\infty$.

1.4 A system design problem

We now formalize the generic problem we attempt to solve in this book.

The designer rejects or accepts a system by comparing the actual characteristics of the system to a set of requirements that describes what is expected from the system. The two main types of requirements we consider are internal stability and performance.

The system $\mathcal{S}$ is *internally stable* if its closed-loop transfer function T is stable and there is no cancellation of left-half-plane (unstable) poles and zeros in the product PC. Internal stability is treated in Chapter 2.

In the simplest possible terms, a performance requirement is an inequality in terms of the designable closed-loop transfer function T. An example is

$$|T(j\omega)| < 0.707 \quad \text{for } \omega > 1.0.$$

Performance requirements are discussed in Chapter 3.

We state the design problem as follows:

---

[2]This has the advantage that design specifications are usually presented in terms of the closed-loop transfer function T. It also has a mathematical advantage that is important to the optimization algorithms underlying our method.

1.5. THE METHOD

Design   Given a plant P, a set of performance requirements $\mathcal{P}$, and a set of feasible functions $\mathcal{I} \subset \mathcal{RH}^\infty$, determine if there exist $T \in \mathcal{I}$ that make the closed-loop system $\mathcal{S}$ internally stable and satisfy all the performance requirements in $\mathcal{P}$. If the answer is yes, find one such T.

1.5 The method

A solution to *Design* is found by the following sequence of steps.

I. Obtain a mathematical description of the performance and internal stability requirements.

Performance requirements. Each performance requirement is an inequality in terms of the designable transfer function T. In Chapter 3 we present a list of frequency domain performance requirements from which the designer can choose to formulate problems.

Internal stability requirements. Designable transfer functions T corresponding to internally stable systems can be parameterized by a formula (see Chapter 2). The beginner who simply wants to run a design package does not need to know much about internal stability in that many design packages do this step automatically.

II. Obtain a performance function and a performance index from the performance function.

The mathematical expressions for the performance requirements are combined to form a *performance function* $\Gamma(\omega, T(j\omega))$, a positive-valued function of T and ω. In many cases the performance function is defined in such a way that a designable T satisfies the requirements if and only if

$$\Gamma(\omega, T(j\omega)) \leq 1 \quad \text{for all } \omega. \tag{1.7}$$

By considering the worst-case performance over all frequencies, a *performance index* that depends on the closed-loop transfer function T is obtained:

$$\gamma(T) = \sup_\omega \Gamma(\omega, T(j\omega)). \tag{1.8}$$

The performance index is a single number that gives a measure of goodness of a choice of T. In other words, it is a cost function: the smaller $\gamma(T)$, the better the choice of T. A procedure to form Γ from the requirements is discussed in Chapters 3 and 4.

III. Minimize the performance index over all T that correspond to an internally stable system $\mathcal{S}$.

This is the step that requires a special-purpose computer optimization program. Minimizing the performance index over all possible T produces two outputs:

- A yes or no answer to the question of whether there exist closed-loop systems that meet all the requirements.

- An optimal designable transfer function T^*.

The function T^* is (among the closed-loop transfer functions that satisfy internal stability requirements) the best, in terms of overall performance as measured by the index. Optimization (minimizing the performance index) is treated in Chapter 4, and a computer session is presented in Chapter 5.

If it is determined in step III that there exist solutions to *Design*, then the optimal designable transfer function T^* can be used to run simulations, can be implemented physically, or, as will often be the case, can serve as a guide to restating the problem with more stringent performance requirements. The new *Design* problem is subsequently solved with steps I, II, and III.

If the result of step III is that no solutions to *Design* exist, the engineer is confronted with two possible paths. One is to redefine the problem completely by radically changing the specs. The other is to reexamine the requirements to determine which of them can be relaxed. In this case, a problem is formulated with the new set of requirements and then solved with steps I, II, and III.

Since in practice it is common to repeat steps I, II, and III several times in the manner described above, we add another step to our list. With step IV one can treat a sequence of *Design* problems.

IV. If many satisfactory closed-loop systems exist (or none exist), tighten (or loosen) the specifications accordingly and go to Step I. Stop if this process can not be carried further.

The mathematics used here produce T^* and an associated C^*, via equation (1.6), given by a set of values on the $j\omega$ axis. In most instances the designer will want to represent C^* as a rational function. In some cases, it is desirable that this rational function have low order. These two objectives can be accomplished with the techniques provided by the subjects of system identification and model reduction. While these subjects are not treated in this book,[3] the software package OPTDesign has functions for doing (stable or otherwise) rational fits of data.

We finish this chapter with a caveat. It may be desirable in doing system design to select compensators that are stable. For example, this would be the case if the engineer wants to build and test the compensator as an independent unit. In its current state, the theory and computational methods of H^∞ optimization do not handle stability of compensators. Thus optimal compensators C^* produced by the methods used here are not necessarily stable (see [DFT92], page 79).

---

[3] The reader is referred to [SS90] and [GKL89] for theoretical aspects of system identification and model reduction. See also [B92] and [Tr86].

1.6 Exercises

1. Verify that the properties of being real, bounded, stable, and proper are preserved under the following operations: addition of two rational functions, multiplication of rational functions, and multiplication of a rational function by a real number.

2. Prove that a rational function is real if and only if the coefficients can be chosen to be real numbers.

3. Prove that a rational function F is bounded if and only if it is proper and has no poles on the imaginary axis.

4. Use formula (1.6) in verifying the following relations:
   a. $\frac{dT}{dP} = S\frac{T}{P}$
   b. $S + T = 1$
   c. $L = \frac{T}{1-T}$

5. Write all functions L, C, S, and Q in terms of P and T.

6. Consider a closed-loop system $\mathcal{S}$ where T is stable and no right-half-plane (RHP) pole-zero cancellation occurs in PC. Prove that in this case
   a. If s_0 is a zero of P in the RHP, then $T(s_0) = 0$.
   b. If s_1 is a pole of P in the RHP, then $T(s_1) = 1$.

7. Rewrite the relation
$$|S(j\omega)| < 0.5 \quad \text{for } 0 \leq \omega < 1$$
in terms of $T(j\omega)$.

Chapter 2

Internal Stability

This chapter is an introduction to internal stability of systems. The main concept is discussed in section 2.1, where it is illustrated with examples. Internal stability is closely connected to a mathematical topic called interpolation, and that is the subject of section 2.2. We conclude the chapter with a section on control systems for a given stable plant. The chapter gives the student a quick overview of internal stability, as well as enough background to treat design problems of certain types. The discussion of internal stability in full generality is left to the more advanced student and is presented in Chapter 7.

2.1 Control and stability

Consider the closed-loop system $\mathcal{S}$ depicted in Fig. 2.1, where P and C are proper, real, rational functions. Given a plant P, by choosing an appropriate compensator C one can obtain a transfer function T for the closed-loop system that is stable, i.e., that has no poles in the closed right-half-plane (RHP).

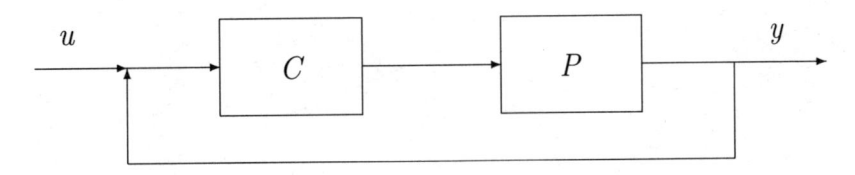

Fig. 2.1. *The closed-loop system $\mathcal{S}$.*

One way to achieve this is to choose C that cancels RHP poles and zeros of P. However, any RHP pole-zero cancellation in the product PC is highly undesirable, because small uncertainties in the plant P or in the compensator's construction lead to a radical change in the behavior of the closed-loop system $\mathcal{S}$. We now illustrate this point with an example.

CHAPTER 2. INTERNAL STABILITY

Example. For the given plant $P(s) = 1/(s-1)$, consider the compensators $C_1(s) = (s-1)/(s+1)$ and $C_2(s) = 2$. Observe that a RHP zero of the compensator C_1 cancels a RHP pole of the plant when the product $P(s)C_1(s)$ is formed:

$$P(s)C_1(s) = \frac{1}{s-1}\frac{s-1}{s+1} = \frac{1}{s+1}. \tag{2.1}$$

That the closed-loop transfer function T_1 associated with C_1 is stable is clear from the following calculation:

$$T_1(s) = \frac{P(s)C_1(s)}{1+P(s)C_1(s)} = \frac{\frac{1}{s+1}}{1+\frac{1}{s+1}} = \frac{1}{s+2}.$$

Now consider a small perturbation of the plant corresponding to a change in the location of the pole. Thus we now have the plant

$$P^\delta(s) = \frac{1}{s-1+\delta}, \tag{2.2}$$

where δ is a real number with small absolute value. For this plant and the compensator C_1, the closed-loop transfer function is

$$T_1^\delta(s) = \frac{P^\delta(s)C_1(s)}{1+P^\delta(s)C_1(s)} = \frac{\frac{s-1}{(s+1)(s-1+\delta)}}{1+\frac{s-1}{(s+1)(s-1+\delta)}} = \frac{s-1}{(s+1)(s-1+\delta)+(s-1)}.$$

The function T_1^δ has an unstable pole for all small values of δ, except for $\delta = 0$. Let s_δ denote such value. Figure 2.2 shows the plot of s_δ. Here

$$s_\delta = \frac{1}{2}\left(-(1+\delta) + \sqrt{(1+\delta)^2 - 4(\delta-2)}\right).$$

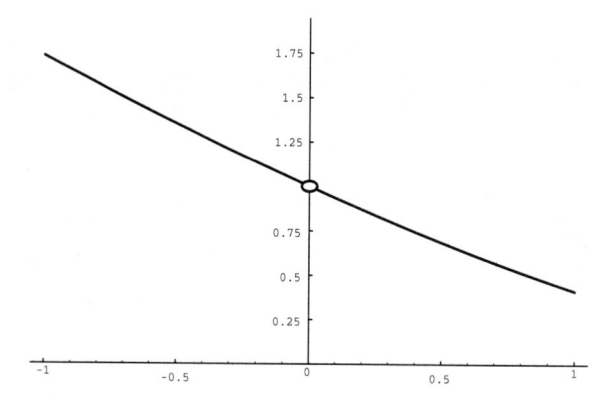

Fig. 2.2. *Plot of the unstable pole s_δ of P^δ as a function of δ. The parameter δ is on the horizontal axis. The point $(0,1)$ is omitted, since for $\delta = 0$ there is no unstable pole in P^δ.*

2.2. INTERPOLATION

On the other hand, for the plant P^δ and compensator C_2 the closed-loop transfer function is

$$T_2^\delta(s) = \frac{P^\delta(s)C_2(s)}{1 + P^\delta(s)C_2(s)} = \frac{\frac{2}{s-1+\delta}}{1 + \frac{2}{s-1+\delta}} = \frac{2}{s+1+\delta}.$$

Note that T_2^δ is stable for all small δ, while T_1^δ is not.

In this example we see that cancellation of a RHP pole of P with a zero of C produces an undesirable system, since the latter becomes unstable under small perturbations of the plant. A similar phenomenon occurs when a RHP zero of the plant is canceled with a pole of the compensator (see Exercise 1). Thus stability of T, sometimes referred to as *external stability* of the closed-loop system, is not sufficient to guarantee a satisfactory closed-loop system $\mathcal{S}$, although it is a necessary ingredient in the practical design of systems.

The closed-loop transfer function describes external, or input-output, behavior of the system. A more satisfying concept of stability is that of *internal stability*.

DEFINITION. *The system $\mathcal{S}$ is* internally stable *if the following are all satisfied.*

i. *The closed-loop transfer function T is in $\mathcal{RH}^\infty$.*

ii. *A RHP pole of the plant is not canceled by a RHP zero of the compensator.*

iii. *A RHP zero of the plant is not canceled by a RHP pole of the compensator.*

Here $\mathcal{RH}^\infty$ is the linear space of all proper rational functions that are stable (poles off the closed *RHP*) and real (real coefficients).

The fundamental problem we treat in this chapter is as follows: given a plant P in a certain class, find a description of *all* the internally stable systems $\mathcal{S}$ with P as plant. The same problem is discussed in Chapter 7 for any plant P.

There are many ways in which one can present an answer to this problem, all of which are mathematically equivalent. In this book we choose one that consists of writing *a formula* that is (and must be) satisfied by all designable closed-loop transfer functions T of these systems. The reasons for choosing this particular approach are technical rather than physical: we have a procedure available that uses this formula to give an answer to system design problems.

To derive formulas for internally stable systems we need to introduce the concept of *interpolation condition*, which is presented in the next section.

2.2 Interpolation

Given s_0, a complex number, and f, a rational function, a relation of the form

$$f(s_0) = v_0 \qquad (2.3)$$

is called an *interpolation condition on f*. Interpolation constraints on functions in $\mathcal{RH}^\infty$ play an important role in H^∞ control. One of the main tools in solving problems involving interpolation constraints is a formula for parameterizing all $\mathcal{RH}^\infty$ functions meeting interpolation conditions.

A case familiar to all is that of polynomial f (rather than rational). A basic result from algebra states that given a point s_0 in the complex plane and a polynomial f, then

$$f(s_0) = 0 \tag{2.4}$$

if and only if one can write

$$f(s) = (s - s_0)h(s), \tag{2.5}$$

where h is some polynomial. This can be thought of as a formula or parameterization of all functions f satisfying (2.4).

To illustrate the role of interpolation in internal stability, we consider the case of a system $\mathcal{S}$ with a plant P that has a first-order (simple) pole at $s = s_0$ in the RHP. In the discussion below the following interpolation condition is important:

$$T(s_0) = 1 \tag{2.6}$$

CLAIM. *If s_0 is a simple RHP pole of P, and T satisfies (2.6), then there is no pole-zero cancellation in the product $P(s)C(s)$ at $s = s_0$.*

Proof. If there is cancellation, then $P(s_0)C(s_0)$ is finite, and this implies that

$$T(s_0) = \frac{P(s_0)C(s_0)}{1 + P(s_0)C(s_0)} \neq 1.$$

This proves the claim.

CLAIM. *If s_0 is a simple RHP pole of P, and if there is no RHP pole-zero cancellation in PC at $s = s_0$, then T satisfies (2.6).*

Proof. If there is no cancellation in the product $P(s)C(s)$ at $s = s_0$, then $P(s)C(s)$ has an order $n_0 \geq 0$ pole at $s = s_0$. Let

$$\ell := \lim_{s \to s_0} (s - s_0)^{n_0} P(s)C(s).$$

Then ℓ is a complex number and

$$\begin{aligned} T(s_0) &= \lim_{s \to s_0} \frac{P(s)C(s)}{1 + P(s)C(s)} \\ &= \lim_{s \to s_0} \frac{(s - s_0)^{n_0} P(s)C(s)}{(s - s_0)^{n_0} + (s - s_0)^{n_0} P(s)C(s)} = \frac{\ell}{0 + \ell} = 1. \end{aligned}$$

The above claims form the first half of the following result.

PROPOSITION 2.2.1. *Let $\mathcal{S}$ be a system with plant P and closed-loop transfer function T.*

2.3. SYSTEMS WITH A STABLE PLANT

- If s_0 is a simple RHP pole of P, then pole-zero cancellation in the product PC does not occur at $s = s_0$ if and only if $T(s_0) = 1$.

- If s_0 is a simple RHP zero of P, then pole-zero cancellation in the product PC does not occur at $s = s_0$ if and only if $T(s_0) = 0$.

One consequence of Proposition 2.2.1 is that for certain systems, internal stability is described in terms of a set $\mathcal{J}$ of interpolation conditions such as $T(s_0) = 0$ or $T(s_0) = 1$. A complete discussion of this can be found in Chapter 7, but in this chapter we restrict our study to the case where the plant is stable.

We shall see in Chapter 7 that one can always write down a formula for all functions T in $\mathcal{RH}^\infty$ that satisfy given interpolation conditions. Such a formula has the form

$$T(s) = A(s) + B(s)\, T_1(s), \tag{2.7}$$

where A and B are certain functions in RH^∞ that can be determined from $\mathcal{J}$, and T_1 could be any function in $\mathcal{RH}^\infty$. Thus as T_1 sweeps $\mathcal{RH}^\infty$, we have T sweeping all functions in $\mathcal{RH}^\infty$ meeting $\mathcal{J}$.

2.3 Systems with a stable plant

Systems $\mathcal{S}$ with a RHP-stable plant are a common type in practical control systems. These are also simple enough to provide a good introduction to the study of internal stability.

We now try to answer the question, given a stable plant P, what systems $\mathcal{S}$ have P as plant and are internally stable?

To give an answer we will find a formula for the closed-loop transfer functions T that come from such systems. Here is a restatement of the question above: given a stable plant P, describe all possible closed-loop transfer functions T that come from an internally stable system $\mathcal{S}$ with plant P. The answer we seek is in the following theorem.

THEOREM 2.3.1. *Let P be a strictly proper, RHP-stable plant for the system $\mathcal{S}$. Then the system $\mathcal{S}$ is internally stable if and only if there exists $T_1 \in \mathcal{RH}^\infty$ such that $T = PT_1$.*

Proof. If $\mathcal{S}$ is internally stable, then PC has the same zeros as P, including multiplicity. Thus the zeros of $T = PC/(1 + PC)$ contain the zeros of P, and this implies that the function $T_1 := T/P$ has no RHP poles. The function T_1 is proper since the system $\mathcal{S}$ has proper C and $T_1 = C/(1 + PC)$. Thus we have that T_1 is an element of $\mathcal{RH}^\infty$. This proves the "only if" side of the theorem. Conversely, suppose that $T = PT_1$ for some $T_1 \in RH^\infty$. Note that the compensator is given by

$$C = \frac{T}{P(1-T)} = \frac{T_1}{1 - PT_1}.$$

It is clear from this formula and from P being strictly proper that C is proper. For C to have a pole at the same location $s = s_0$ as a RHP zero of P, one must have $1 - P(s_0)T_1(s_0) = 0$, which is impossible since T_1 has no pole at s_0 and P is zero there. ∎

Example. If the plant of an internally stable system is $P(s) = \frac{1}{(s+3.4)^2}$, then any stable closed-loop transfer function T has the form

$$T(s) = \frac{1}{(s+3.4)^2} T_1(s), \tag{2.8}$$

where T_1 is some (or any) element of $\mathcal{RH}^\infty$. We note that T_1 may not be strictly proper. In fact, $d(T_1) \neq 0$ forces $d(C) = 0$ upon the system.

Note that we used the actual plant P in the formula (2.8). While this may be convenient, many other formulas for T are possible. For example, we can correctly write

$$T(s) = \frac{1}{(s+1)^2} T_1(s) \quad \forall T_1 \in \mathcal{RH}^\infty \tag{2.9}$$

to describe all T from internally stable $\mathcal{S}$.

Example. If now $P(s) = (s^2 - 4)/(s^4 + 2s^2 + 2)$, then P is stable and strictly proper, so for the system to be internally stable the closed-loop transfer function must have the form

$$T(s) = \frac{s^2 - 4}{s^4 + 2s^2 + 2} T_1(s)$$

for some $T_1 \in \mathcal{RH}^\infty$.

2.4 Exercise

1. Consider the family of plants given by

   $$P^\delta(s) = (s - 1 + \delta)/(s + 1),$$

   where δ is a real constant. Show that

   a. For small $|\delta|$, the compensator $C_1(s) = 1/(s-1)$ and plant P^δ yields a closed-loop transfer function T^δ that is stable if and only if $\delta = 0$.

   b. For all small $|\delta|$, the compensator $C_2(s) = (s+1)/(s+3)$ and plant P^δ yields a stable closed-loop transfer function T^δ.

Chapter 3

Frequency Domain Performance Requirements

Frequency domain performance requirements have a convenient graphical interpretation in terms of disks. Section 3.1 reviews basic concepts and introduces disk inequalities. The most common performance requirements for control systems are given in section 3.2. Section 3.3 discusses disk inequalities arising from performance requirements. The first-time reader may stop after reading this section and jump to Chapter 4. The topics covered at this point are sufficient to provide the basic tools to solve simple design problems, when used in conjunction with material from Chapters 2, 4, and 5. The more interested reader may continue with sections 3.4 and 3.5, which explain additional measures of performance and the corresponding disk inequalities.

3.1 Introduction

3.1.1 The closed-loop system $\mathcal{S}$

Consider the closed-loop system depicted in Fig. 3.1. In this system, $P(s)$ and $C(s)$ are proper real rational functions of s. We take the plant P to be a given rational function and use the closed-loop transfer function T to parameterize the closed-loop systems $\mathcal{S}$ obtained with different compensators. Hence many systems $\mathcal{S}$ with given plant P are possible by letting the designable transfer function T take different values. Key functions are

- The closed-loop function $T = \frac{PC}{1+PC}$;
- The sensitivity function $S = (1 + PC)^{-1}$;
- The tracking error for input u, Su;
- The closed-loop compensator $Q := CS$;

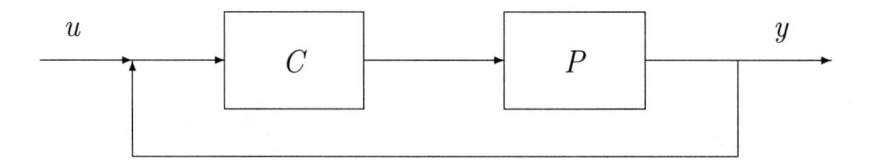

Fig. 3.1. *The closed-loop system* $\mathcal{S}$.

- The closed-loop plant PS;

- The open-loop transfer function $L := PC$.

3.1.2 Frequency domain performance requirements

Two types of requirements imposed on the system $\mathcal{S}$ are discussed in this book: internal stability and performance. The internal stability requirement was studied in Chapter 2. Performance requirements are inequalities involving the functions of the system. In this chapter we focus on performance requirements in the frequency domain.

DEFINITION. A *frequency domain performance requirement* is an inequality that the designable transfer function T must satisfy, and an interval in ω for which the inequality is required to hold.

3.1.3 Disk inequalities

All of the frequency domain performance requirements treated in this chapter can be written in the form of a *disk inequality*

$$|K(j\omega) - T(j\omega)| \leq R(j\omega), \quad \omega_a \leq \omega \leq \omega_b, \qquad (3.1)$$

which T must satisfy. Here K and R are fixed functions that embody the desired specs of the system. K is called the *center* of the disk (3.1) and R is called the radius of (3.1). Disk inequalities are easy to plot as regions in 3-D space (see Fig. 3.2) and correspond to the inside of a tubelike domain. If the frequency ω is fixed, then a disk inequality is represented in the complex plane by a region

$$S_\omega = \{z : |K(j\omega) - z| \leq R(j\omega)\}$$

consisting of a solid disk.

In many cases the requirements give one disk inequality on one frequency band, and other disk inequalities on other frequency bands. When the various frequency bands are disjoint, it is possible to link these disk inequalities together into a single one, valid for all frequencies:

$$|K(j\omega) - T(j\omega)| \leq R(j\omega) \quad \text{for all } \omega. \qquad (3.2)$$

3.2. MEASURES OF PERFORMANCE

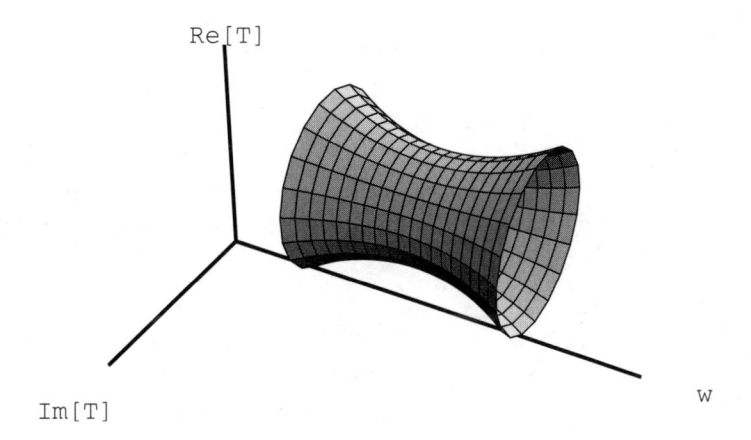

Fig. 3.2. *Tubelike region $\{(\omega, z) : |K(j\omega) - z| = R(j\omega)\}$ determined by a disk inequality $|K(j\omega) - T(j\omega)| \leq R(j\omega)$.*

The process of piecing together disk inequalities will be discussed in sections 3.3 and 3.5. We shall see in Chapter 4 how to use disk inequalities to pose optimization problems of the simplest kind discussed in this book. Many system design problems can be solved by finding solutions to these optimization problems.

If two or more disk inequalities apply at the same frequency ω, then the region S_ω is the intersection of the disks that corresponds to individual constraints. Thus in this case, S_ω is not a disk.

3.2 Measures of performance

A set of basic frequency domain performance requirements are presented in this section. More requirements are introduced later in the chapter; however, it is possible to state physically sensible design problems with the requirements given in this section alone.

3.2.1 Gain-phase margin

In classical control, the gain margin and phase margin are the chief measures of how robustly stability is achieved. In order to formulate constraints in terms of the closed-loop transfer function T, we use a composite measure of stability, which seems at least as good as the gain margin and phase margin taken together. Define the *gain-phase margin* of the system S to be

$$m = \inf_\omega |1 + PC(j\omega)|. \tag{3.3}$$

Graphically m is just the distance of the Nyquist plot of PC to the point -1 in the complex plane. Simple algebra converts (3.3) to

$$|T(j\omega) - 1| \leq \frac{1}{m} \quad \text{for all} \quad \omega. \tag{3.4}$$

We can easily compare the gain-phase margin m with the gain margin g and the phase margin ϕ by looking at the Nyquist plot in Fig. 3.3. Typically m is more conservative than either ϕ or g.

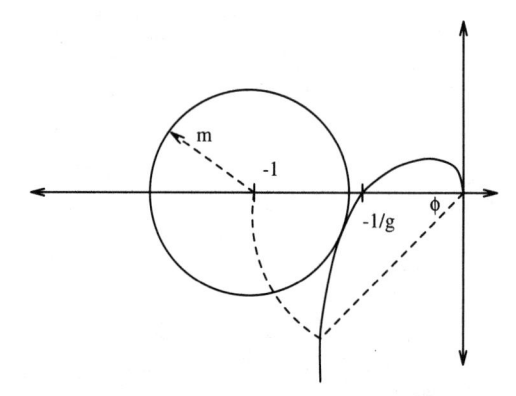

Fig. 3.3. *The gain margin g, phase margin ϕ, and gain-phase margin m.*

If a given stable system has m near 0, then it is close to the unstable case, which is undesirable. If α_m denotes the largest value of $1/m$ considered to be acceptable in inequality (3.4), then we can formulate a constraint in terms of α_m as (see Fig. 3.4)

$$|1 - T(j\omega)| \leq \alpha_m \quad \text{for all} \quad \omega. \tag{3.5}$$

Fancier notions are very natural. For example, if W is a given nonnegative function of frequency, we can define

$$m_W \triangleq \inf_\omega |W(j\omega)(1 + PC)(j\omega)| \tag{3.6}$$

to be the *weighted gain-phase margin*. The corresponding constraint is derived from this definition in the obvious way and produces $\alpha_m(j\omega) = W(j\omega)/m_W$.

Gain-phase margin constraint

For a given α_m in the interval $(0, 1)$,

$$|1 - T(j\omega)| \leq \alpha_m \quad \text{for all} \quad \omega. \tag{3.7}$$

3.2. MEASURES OF PERFORMANCE

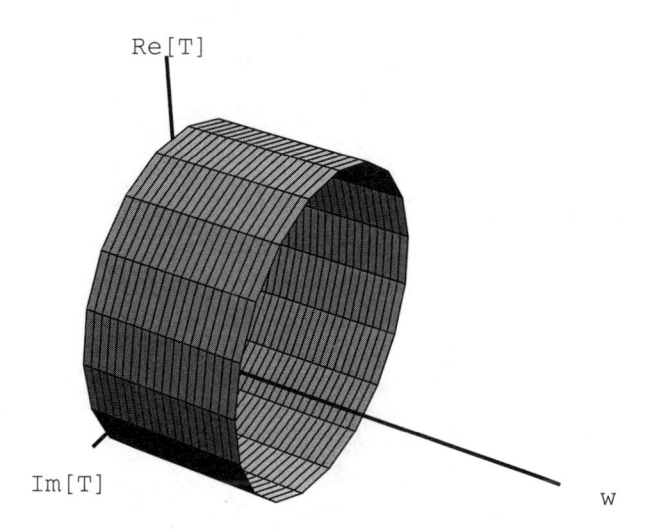

Fig. 3.4. *Region defined by the gain-phase margin constraint.*

3.2.2 Tracking error

Fundamentally, tracking error is a time domain concept. The issue is to measure how close the output $y(t)$ of the closed-loop system is to a given input $u(t)$. Indeed, we want the error function

$$\epsilon_u(t) = u(t) - y_u(t) \tag{3.8}$$

to be "small" for each function u in a big class of possible inputs. To use frequency domain techniques, one must translate time domain criteria back to the frequency domain. We do this below.

Consider system $\mathcal{S}$ with given time domain input $u(t)$ and output $y(t)$, and let $U(s)$ and $Y(s)$ be the Laplace transforms of $u(t)$ and $y(t)$. Then $Y(s) = T(s)U(s)$, and the Laplace transform of ϵ_u is

$$E_u(s) = (1 - T(s))U(s). \tag{3.9}$$

For $T \in \mathcal{RH}^\infty$, a condition that generates small ϵ_u compared to the size of u is

$$\sup_\omega |1 - T(j\omega)| < \alpha_{tr} \tag{3.10}$$

for some small α_{tr}. Thus good tracking is generated by requiring (3.10) to hold.[1]

It turns out that (3.10) is too stringent to be obtainable in a practical engineering problem since the function T rolls off at high frequency. What is realistic is to require the system to track low-frequency functions well, that is, to require that $T(j\omega)$ be close to 1 over some specified frequency range $[-\omega_{tr}, \omega_{tr}]$. Thus we introduce one of the key constraints of control.

---

[1] A standard theorem (the Plancherel theorem, [Yng88]) applied to (3.9) yields immediately

22   CHAPTER 3.   FREQUENCY DOMAIN PERFORMANCE

Tracking Error Constraint

For given $\alpha_{tr} > 0$, $\omega_{tr} > 0$,

$$|1 - T(j\omega)| < \alpha_{tr} \quad (|\omega| < \omega_{tr}). \tag{3.14}$$

(See Fig. 3.5.)

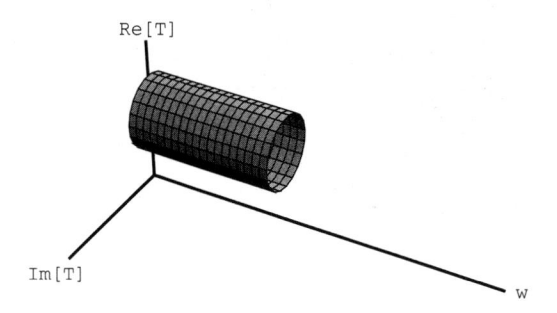

Fig. 3.5. *Region defined by the tracking error constraint.*

3.2.3 Bandwidth

Bandwidth of a system is commonly defined as the frequency ω_b at which $|T(j\omega)|$ falls below a given constant times the low-frequency value of the input. This constant usually is taken to be $\alpha_b = \frac{1}{\sqrt{2}}$. A particular bandwidth is required to ensure that the system at high frequency is not upset by noise, plant uncertainty, actuator sluggishness, etc.

---

that for $T \in RH^\infty$ the error satisfies a useful bound:

$$\begin{aligned}\int_0^\infty |\epsilon_u(t)|^2 dt &= \int_{-\infty}^\infty |E_u(j\omega)|^2 d\omega = \int_{-\infty}^\infty |1 - T(j\omega)|^2 |U(j\omega)|^2 d\omega \\ &\leq \left\{\sup_\omega |1 - T(j\omega)|^2\right\} \int_{-\infty}^\infty |U(j\omega)|^2 d\omega \\ &= \left\{\sup_\omega |1 - T(j\omega)|^2\right\} \int_0^\infty |u(t)|^2 dt \end{aligned} \tag{3.11}$$

for all functions u on $[0, \infty)$ having finite *energy* $\int_0^\infty |u(t)|^2 dt$. The inequality is sharp in the sense that there are input functions that make this inequality as close as one desires to equality. That is,

$$\sup_u \frac{\int_0^\infty |\epsilon_u(t)|^2 dt}{\int_0^\infty |u(t)|^2 dt} = \sup_\omega |1 - T(j\omega)|^2. \tag{3.12}$$

Consequently, if

$$\sup_\omega |1 - T(j\omega)|^2 \leq c \tag{3.13}$$

then the normalized error in the left-hand side of (3.12) is no larger than c. Thus specs of the form (3.13) guarantee that

$$\text{energy}(\epsilon_u) \leq c \, \text{energy}(u)$$

for all u with finite energy. This is a very strong form of tracking.

3.2. MEASURES OF PERFORMANCE

Bandwidth Constraint

Given $\alpha_b > 0$, $\omega_b > 0$,

$$|T(j\omega)| < \alpha_b \text{ for } |\omega| > \omega_b . \tag{3.15}$$

(See Fig. 3.6.)

In practice a more refined constraint is required for very high frequency, since it is desirable that T roll off to 0. This is discussed in section 3.2.4.

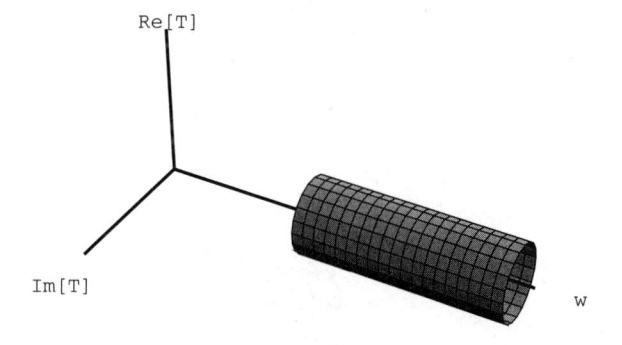

Fig. 3.6. *Region defined by the bandwidth constraint.*

3.2.4 Closed-loop roll-off

The open-loop transfer function $L = PC$ of a given system rolls off at high frequency, since the plant P and the compensator C are strictly proper. That is, $|P(j\omega)C(j\omega)| \to 0$ as $\omega \to \infty$.

From the relation

$$|T(j\omega)| \leq \frac{|P(j\omega)C(j\omega)|}{1 - |P(j\omega)C(j\omega)|} \tag{3.16}$$

$$\approx |P(j\omega)C(j\omega)| \text{ (large } |\omega|), \tag{3.17}$$

we see that both T and PC roll off at the same rate. To obtain an inequality useful for the design process, we must eliminate the compensator C from the right-hand side of relation (3.17). To do this, use the fact that compensators roll off at high frequency with the asymptotic form[2]

$$C(j\omega) \leq \frac{\alpha_r}{|\omega|^n} \tag{3.18}$$

---

[2] Since behavior near $j\infty$ is what is important in (3.18), α_r can be taken to be a weight function of frequency that is not too close to 0 at any given frequency.

for some α_r and some n. At the outset of design, an engineer should specify n and α_r for the class of compensators that is to be built.[3] Once α_r and n are specified, combine (3.17) and (3.18) to obtain the closed-loop roll-off constraint.

Closed-loop roll-off constraint

For given $\omega_r > 0$, $\alpha_r > 0$, and $n > 0$,

$$|T(j\omega)| \leq \frac{\alpha_r |P(j\omega)|}{|\omega|^n} \quad (|\omega| > \omega_r). \qquad (3.19)$$

(See Fig. 3.7.)

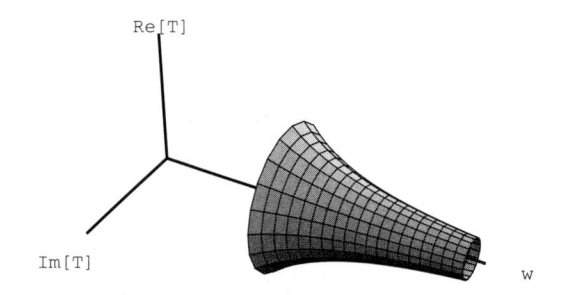

Fig. 3.7. *Region defined by the closed-loop roll-off constraint.*

3.2.5 Fundamental trade-offs

The most fundamental trade-off in control is between bandwidth constraints and performance measures such as tracking or gain-phase margin, which dominate considerations at low frequency. Bandwidth and roll-off are consistent constraints, and tracking error competes strongly with them. Thus a basic challenge in control is to get tracking over a broad enough frequency band, subject to the roll-off constraints dictated by actuator sluggishness and uncertainty in plants, sensors, and the environment. In a particular engineering problem, to obtain a precise feel for this trade-off requires the use of a computer program.

3.2.6 Choosing sets of performance requirements

The designer must choose a set of performance requirements in order to state and then solve a design problem. This set should be selected with care, since a

---

[3] We thank L. Desoer for emphasizing this constraint to us.

bad choice leads to problems that do not make sense from either the numerical or the physical point of view. We illustrate this with examples below.

Example. Consider the problem *Design* for the plant $P(s) = 1/(1-s)$ and the performance requirement

$$|T(j\omega)| < 0.707 \quad \text{for } \omega > 1.0. \tag{3.20}$$

An obvious mistake here is the absence of constraints valid on the band $0 \leq \omega \leq 1.0$. It is easy to find functions T that satisfy (3.20) and that produce internally stable systems $\mathcal{S}$, but that have undesirable behavior at low frequency.

Example. Now consider *Design* for $P(s)$ with performance requirements

$$\begin{array}{rcll} |T(j\omega)| & < & 0.707, & \text{for } \omega > 1.0 \\ |1 - T(j\omega)| & < & 2.0, & \text{for } 0 \leq \omega \leq 1.0. \end{array} \tag{3.21}$$

For any physical system, T must roll off to 0 at very high frequency. This is not enforced by the constraints (3.21). One way to remedy this is to require roll-off on T, for example, with the constraint

$$|T(j\omega)| < \alpha_c |P(j\omega)|, \quad \text{for } \omega > \omega_c.$$

The examples above illustrate two basic principles for choosing a set of frequency domain performance requirements:

- At each frequency ω, there must be a constraint in the set of requirements that is active at this frequency.

- Behavior of the system at very high frequency must be specified with a roll-off constraint on T.

There are cases where performance measures not described in this section are important. In particular, if the plant has a zero or a pole right on the $j\omega$ axis or near it, the *compensator bound* and *plant bound* constraints must be used to set up the problem correctly. See section 3.4 for details on this.

3.3 Piecing together disk inequalities

The performance measures introduced in section 3.2 can be expressed as *disk inequalities*. In this section, we present an example in which several disk inequalities are put together and expressed as one disk inequality that is valid for all frequencies. Being able to combine disk inequalities into a single one is a necessary skill for students of control who want to solve system design problems with the methods proposed in this book.

Example. Consider a case where the requirements are given by constraints only on the tracking error, gain-phase margin, and bandwidth.[4] We will formulate a single disk constraint from the given constraints.

Suppose the following constraints are given:

---

[4]This is not a physical example, since there is no closed-loop roll-off requirement.

Table 3.1. *A design example with three constraints.*

Constraint	Disk Inequality	Freq. Band	$K(j\omega)$	$R(j\omega)$
Tracking and disturbance rejection	$\|T(j\omega) - 1\| < 0.25$	$0 \leq \omega < 0.5$	1	0.25
Gain-phase	$\|T(j\omega) - 1\| < 2.0$	$0.5 \leq \omega < 3.0$	1	2.0
Bandwidth	$\|T(j\omega)\| < 0.707$	$3.0 \leq \omega < \infty$	0	0.707

- $|1 - T(j\omega)| < 0.25$ for $\omega < 0.5$ (tracking for unit step input),
- $|1 - T(j\omega)| \leq 2.0$ for all ω (gain-phase margin),
- $|T(j\omega)| \leq 0.707$ for $|\omega| > 3$ (bandwidth).

Note that for ω fixed, the large disks from the gain-phase margin constraint contain the smaller disks from the bandwidth and the tracking error constraints. Thus the smaller disks define the region where $T(j\omega)$ must lie for either low or high frequencies ω. For midrange frequencies, the gain-phase margin constraint is the only constraint that is applicable. We collect this information in Table 3.1. The notation there is the same as the notation in inequality (3.1). The region defined by these constraints is drawn in Fig. 3.8.

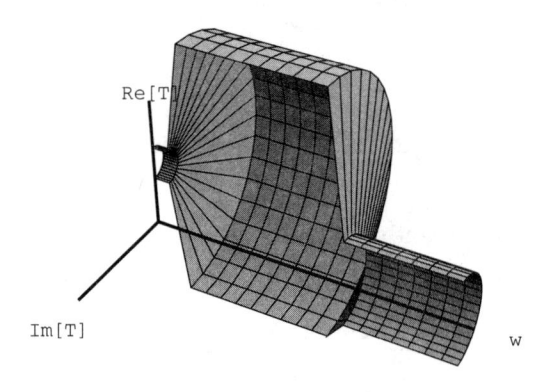

Fig. 3.8. *The region defined by the constraints in Table* 3.1.

The point is that at each frequency ω there is one and only one disk

$$\mathcal{S}_\omega := \{z \; : \; |k(j\omega) - z| < r(j\omega)\}$$

in which $T(j\omega)$ is constrained to lie. This is a mathematical description of performance specifications that can easily be put in a computer program of the kind described in this book.

3.4 More performance measures

In section 3.2 we showed how to convert the most basic specifications in a control problem to precise performance measures (suitable for computer optimization). A beginner who masters these alone will have a very good feel for the basics of control.[5] However, there are common control problems that cannot be formulated correctly using only the constraints in section 3.2 — for example, if the plant has a pole or a zero on the $j\omega$ axis. In this section we give additional performance measures, in particular ones that deal with zeros or poles of the plant on the $j\omega$ axis.

3.4.1 Peak magnitude

One common constraint is that the magnitude of the closed-loop transfer function T not become too large; if it does, then T is close to unstable. Also, it has been seen in practical design (see [Ho63], page 190) that the peak magnitude of T is related to a large overshoot in the step response function. In many cases a peak value of 1.1–1.5 is acceptable, while in others this is too high. We now define the peak magnitude constraint; the corresponding graphical representation is given in Fig. 3.9.

Peak magnitude constraint

For given $\alpha_M > 0$,

$$|T(j\omega)| \leq \alpha_M \quad (\text{all } \omega). \tag{3.22}$$

3.4.2 Compensator bound

Desoer and Gustafson proposed another very sensible criterion for good compensation in [DG82]:

$$|C(j\omega)(1 - T(j\omega))| \leq \alpha_c \quad (\text{all } \omega) \tag{3.23}$$

for some prescribed number α_c. In terms of the closed-loop system $\mathcal{S}$, this says that the "closed-loop compensator" has magnitude bounded by α_c. The *closed-loop compensator* is what the compensator puts out in response to an

---

[5]Also, the examples in Chapters 4 and 5 will be clear to this beginner.

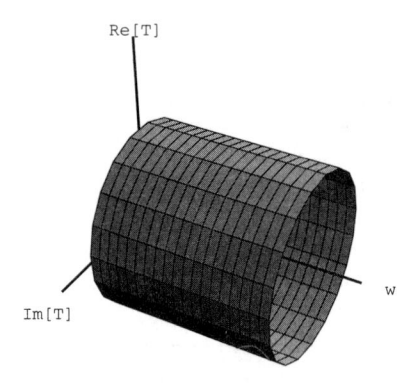

Fig. 3.9. *Region defined by the peak magnitude constraint.*

input to the system. The reason that (3.23) is imposed on the system is that if it is violated, then the closed-loop system can saturate, a small current into the system can cause arcing at the output of C, or other problems may appear. In [DS81] Doyle and Stein discuss drawbacks of high loop gain in somewhat different terms. Their main concern is that (3.23) holds at high frequencies; low frequencies are less important.

Let us analyze (3.23) in terms of T. Since $PC = T/(1-T)$, we obtain

$$|T(j\omega)| \leq \alpha_c |P(j\omega)| \quad \text{(all } \omega\text{)}. \tag{3.24}$$

Typically this constraint adds no serious restriction unless $|P|$ is small (e.g., as $\omega \to \infty$ or near zeros of P on the $j\omega$ axis). Inequality (3.24) adds a restriction that is binding near the zeros of P ($\omega = \infty$ included). Thus the designer should require (3.24) to hold whenever $|\omega - \omega_z| < \eta$, where $j\omega_z$ is any zero of P on the axis and η is a small positive number. Note that for large ω, inequality (3.24) is contained in the closed-loop roll-off constraint (set $n = 0$ in (3.19)).

Compensator bound constraint

For given $\alpha_c > 0$, η_z, and for any zero $j\omega_z$ of P on the $j\omega$ axis,

$$|T(j\omega)| < \alpha_c |P(j\omega)| \quad \text{whenever } |\omega - \omega_z| < \eta_z. \tag{3.25}$$

(See Fig. 3.10.)

3.4.3 Plant bound

Another constraint is that the "closed-loop plant" be bounded by a specified value α_p. That is,

$$|(1 - T(j\omega))P(j\omega)| \leq \alpha_p \quad \text{for all } \omega. \tag{3.26}$$

3.4. MORE PERFORMANCE MEASURES

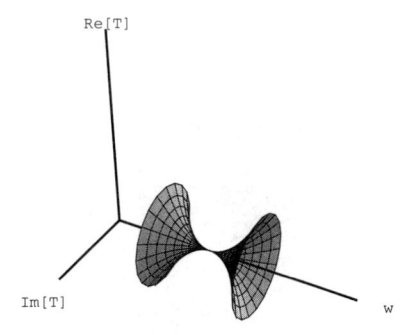

Fig. 3.10. *Region defined by the compensator bound constraint, if there is a zero of P at $s = j\omega_z$.*

The closed-loop plant is the output of the closed-loop system of Fig. 3.1 to an input to the plant.

Inequality (3.26) is analyzed in a similar fashion as inequality (3.24). Recall that internal stability of the system implies that $T(j\omega) = 1$ when $P(j\omega) = \infty$. Inequality (3.26) is binding near the poles of P; it must hold whenever $|\omega - \omega_p| < \eta_p$, where ω_p is any pole of P on the $j\omega$ axis and η_p is a small positive number. From this we get the plant bound constraint, illustrated in Fig. 3.11.

Plant bound constraint

For given $\alpha_p > 0$ and $\eta_p > 0$ and all poles $j\omega_p$ of $P(j\omega)$,

$$|1 - T(j\omega)| < \frac{\alpha_p}{|P(j\omega)|} \quad \text{whenever} \quad |\omega - \omega_p| < \eta_p . \qquad (3.27)$$

3.4.4 Disturbance rejection

Systems are affected by disturbances, and one of the purposes of feedback control is to minimize the effect of these disturbances. As an example, consider the system given by an airplane, with input (established by the pilot) a certain yaw rate $u(t)$ and output the actual yaw rate of the airplane $y(t)$. A gust of wind produces yaw and thus affects the output of the system. A simple way of modeling this situation is shown in Fig. 3.12, where the disturbance is represented as an additive input to the system at the output to the plant. If there is no other input to the system in Fig. 3.12, the transfer function from disturbance d to plant output y is easily found to be

$$T_{d \to y} = \frac{1}{1 + P(s)C(s)}. \qquad (3.28)$$

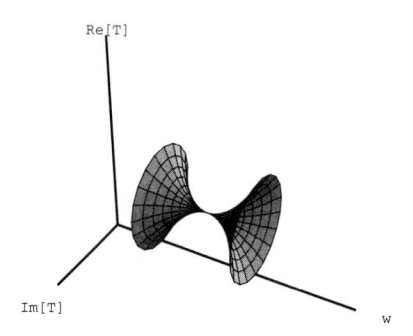

Fig. 3.11. *Region defined by the plant bound constraint, if P has a pole at* $s = j\omega_p$.

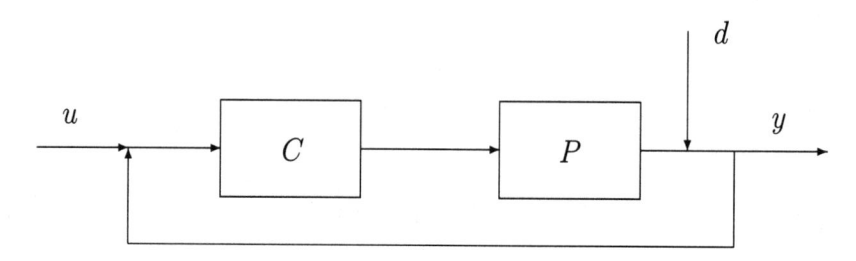

Fig. 3.12. *Disturbance in a feedback system.*

For good behavior of the feedback system it is necessary that $T_{d \to y}$ be RHP-stable. This is always the case with internally stable systems. More can be required — for example, that there is a restriction on the size of $T_{d \to y}$. The following inequality gives a precise statement, where c is a given constant:

$$\left| \frac{1}{1 + P(j\omega)C(j\omega)} \right| \leq c \quad \text{for all } \omega. \tag{3.29}$$

We have already encountered inequalities similar to (3.29) before in the context of tracking error constraint, so at this point we refer the reader to the discussions of the latter.

3.4.5 More on tracking error

Consider a system $\mathcal{S}$ with input u and output y. The steady-state error for input $u(t)$ is

$$e_{ss} = \lim_{t \to \infty} (u(t) - y(t)). \tag{3.30}$$

In many cases relation (3.30) has a counterpart in the frequency domain, which can be obtained with the final value theorem (cf. [C44], page 191, or [LP61], page 315). Suppose that $u(t)$ is such that $\mathcal{L}(u-y)(s)$ has no poles on the closed RHP, except perhaps a simple pole at $s = 0$. The final value theorem says that

3.4. MORE PERFORMANCE MEASURES

in this case
$$\lim_{t\to\infty} u(t) - y(t) = \lim_{s\to 0} s(U(s) - Y(s)).$$

Thus we have from (3.30) that
$$e_{ss} = \lim_{s\to 0} s(U(s) - Y(s)). \tag{3.31}$$

Since $Y(s) = T(s)U(s)$, it follows from (3.31) that
$$e_{ss} = \lim_{s\to 0}(1 - T(s))sU(s). \tag{3.32}$$

If α_{tr} denotes the largest acceptable value of e_{ss} in (3.32), then
$$|e_{ss}| = |\lim_{s\to 0}(1 - T(s))sU(s)| \leq \alpha_{tr}. \tag{3.33}$$

If α_{tr} is not 0, one can write an inequality in the variable ω that guarantees (3.33). We call this inequality the tracking error constraint for input $U(s)$. It is defined for those $U(s)$ for which $(1 - T(s))U(s)$ has no poles in the closed RHP, except possibly for a simple pole at $s = 0$.

Tracking error constraint for input U

For given $\alpha_{tr} > 0$, $\omega_{tr} > 0$,
$$|(1 - T(j\omega))(j\omega)U(j\omega)| < \alpha_{tr} \quad (|\omega| < \omega_{tr}). \tag{3.34}$$

3.4.6 Tracking and type n plants

We now discuss the case where the plant P has n pure integrations (a pole of order n at $s = 0$). In this case we say that the plant is of "type n."

We assume that internal stability is a requirement, so RHP poles of the plant P are not canceled by zeros of the compensator C. Since the plant is of type n, the product PC also has a pole of order n at $s = 0$. Hence the sensitivity function $S = (1 + PC)^{-1}$ has a zero of order n at $s = 0$. In particular, we have
$$S(0) = S'(0) = \cdots = S^{(n-1)}(0) = 0. \tag{3.35}$$

It follows from equation (3.35) and from $T = 1 - S$ that T satisfies the interpolation conditions
$$T(0) = 1, \; T'(0) = 0, \ldots, T^{(n-1)}(0) = 0. \tag{3.36}$$

Let us now take $m \leq n + 1$ and $U(s) = 1/s^m$. Because of (3.36), the function
$$(1 - T(s))sU(s) = (1 - T(s))s^{1-m} \tag{3.37}$$

has no poles on the closed RHP, except possibly a pole of order 1 at $s = 0$. Therefore the final value theorem applies, and one has

$$e_{ss} = \lim_{s \to 0}(1 - T(s))s^{1-m}. \tag{3.38}$$

By combining (3.36) with (3.38), and applying l'Hôpital's rule for limits, we obtain the following proposition.

PROPOSITION 3.4.1. *An internally stable system with type n plant produces a steady-state error e_{ss} for input $U(s) = 1/s^m$ given by*

$$e_{ss} = \begin{cases} 1 - T(0) & \text{if } m = 1, \\ -\frac{T^{(m-1)}(0)}{(m-1)!} & \text{if } 1 < m \leq n+1. \end{cases} \tag{3.39}$$

Hence $e_{ss} = 0$ if $1 \leq m \leq n$.

Proposition 3.4.1 implies that for internally stable systems with type n plant, the most appropriate input $U(s) = 1/s^m$ in the constraint (3.34) is $U(s) = 1/s^{n+1}$. Other inputs U that may be suitable are functions that are stable on the closed RHP, except for a pole of order $n+1$ at $s = 0$.

Common time domain specifications on systems with type n plant are on the position, velocity, and acceleration error constants K_0, K_1, and K_2,[6] where

$$K_m = \lim_{s \to 0} s^m P(s) C(s). \tag{3.40}$$

One can show that an internally stable system $\mathcal{S}$ with type n plant satisfies $K_0 = T(0)/(1 - T(0))$ and $K_m = \frac{-m!}{T^{(m)}(0)}$ if $1 \leq m \leq n$.

3.5 A fully constrained problem

In section 3.3 we obtained a single disk inequality from performance requirements on bandwidth, tracking, and gain-phase margin. We consider here a similar problem, but now with requirements on bandwidth, closed-loop roll-off, plant bound, tracking, and gain-phase margin. Again, we will derive a single disk inequality from the requirements.

The given plant is

$$P(s) = \frac{1 - s/5}{s(1 - s)}$$

with requirements

a. $|1 - T(jw)| < 0.3$, for $0 \leq \omega < 1.0$

b. $|1 - T(jw)| < 1/|P(jw)|$, for $0 \leq \omega < 1.0$

c. $|1 - T(jw)| < 2$, for all ω

---

[6]These are denoted by K_p, K_v, and K_a in most books.

3.5. A FULLY CONSTRAINED PROBLEM

Table 3.2. *A design example with five constraints.*

Constraint	Disk Inequality	Freq. Band	$K(j\omega)$	$R(j\omega)$
Plant	$\|T(j\omega) - 1\| < \frac{1}{\|P(j\omega)\|}$	$0 \leq \omega < 0.29$	1	$\frac{1}{P(j\omega)}$
Tracking and disturbance rejection	$\|T(j\omega) - 1\| < 0.3$	$0.29 \leq \omega < 1.0$	1	0.3
Gain-phase	$\|T(j\omega) - 1\| < 2.0$	$1.0 \leq \omega < 2.0$	1	2.0
Bandwidth	$\|T(j\omega)\| < 0.707$	$2.0 \leq \omega < 2.41$	0	0.707
Roll-off	$\|T(j\omega)\| < 4\|P(j\omega)\|$	$2.41 \leq \omega < \infty$	0	$4\|P(j\omega)\|$

  d. $|T(j\omega)| < 0.707$, for $\omega > 2$

  e. $|T(j\omega)| < 4|P(j\omega)|$, for $\omega > 2$

The constraints a, b, and c are active at all frequencies $0 \leq \omega < 1$. Clearly constraint c is contained in constraint a, so we restrict attention to a and b. Solve the equation
$$\frac{1}{|P(j\omega)|} = 0.3$$
to obtain the solution $\omega_0 \approx 0.29$. Thus b is more stringent than a on the frequency band $0 \leq \omega \leq 0.29$, and b is less stringent than a on the band $0.29 \leq \omega < 1$. We see that c is the only constraint that applies on the band $1 \leq \omega \leq 2$. For $\omega \geq 2$ it is clear that every T that satisfies d also satisfies c, so it is enough to study constraints d and e on this band. We solve the equation
$$0.707 = 4|P(j\omega)|$$
and obtain the solution $\omega_1 \approx 2.41$. Therefore constraint d is more stringent on the band $2 \leq \omega < 2.41$, whereas e is more stringent on $2.41 < \omega$. We summarize these findings in Table 3.2, which defines the center function K and the radius function R at each frequency. See Fig. 3.13 for a graphical representation of the functions K and R.

The format of Table 3.2 is convenient for representing constraints in a way that is easy to read. Note that in this example we can write the constraints as a single disk inequality. This is possible here because overlapping constraints can be written as a single disk constraint. However, sometimes it is not possible to define a radius and center functions that incorporate all the information

34   CHAPTER 3. FREQUENCY DOMAIN PERFORMANCE

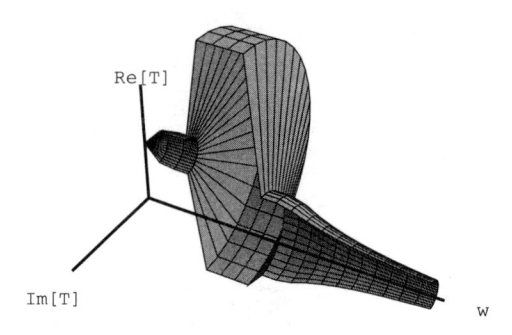

Fig. 3.13. *Section of the envelope defined by the center $K(j\omega)$ and radius $R(j\omega)$ in Table 3.2.*

contained in the requirements. For example, two intersecting disks do not yield a single disk.

Chapter 4

Optimization

In this chapter we discuss performance functions and how to use them as objective functions in optimization problems. These concepts are applied to system design in examples. We begin with a review. The reader who is interested in getting quickly to the point where examples can be solved by computer may read sections 4.1–4.4 and then skip to Chapter 5.

4.1 Review of concepts

Consider again the closed-loop system $\mathcal{S}$ depicted in Fig. 4.1. Here $P(s)$ and $C(s)$ are proper, real, rational functions of s. The function $P(s)$ is given, and the function $C(s)$ depends on the choice of designable transfer function $T(s)$. The functions $T(s)$ and $C(s)$ are related by the formula

$$T(s) = \frac{P(s)C(s)}{1 + P(s)C(s)}$$

or, equivalently,

$$P(s)C(s) = \frac{T(s)}{1 - T(s)}.$$

The system $\mathcal{S}$ is internally stable if T has no RHP poles and there is no RHP pole-zero cancellation in the product PC.

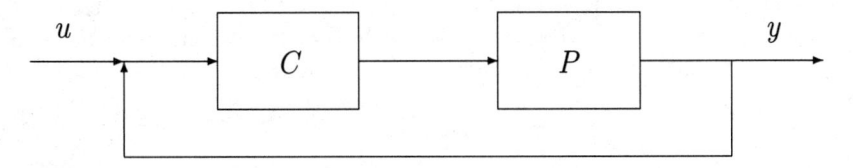

Fig. 4.1. The closed-loop system $\mathcal{S}$.

Performance requirements are given inequalities to be satisfied by functions associated with the system $\mathcal{S}$. The inequalities may be stated in terms of time or frequency. Frequency domain performance requirements were dealt with in Chapter 3. Time domain performance requirements are briefly discussed in Chapter 5 and treated in more detail in Chapter 6.

The basic control problem we consider is

Design Given a plant P, a set of performance requirements $\mathcal{P}$, and a set of feasible functions

$$\mathcal{I} = \{T \in \mathcal{RH}^\infty \ : \ T \text{ is the closed loop of an internally stable system } \mathcal{S} \text{ with plant } P\},$$

determine if there exist $T \in \mathcal{I}$ that satisfy all the performance requirements in $\mathcal{P}$. If the answer is yes, find one such T.

The method used to solve *Design* is as follows:

I. Obtain a mathematical description of the performance and internal stability requirements (write down disk inequalities, and take notice of the unstable poles and zeros of the plant).

II. Obtain a performance function (e.g., derive a formula for the radius and center functions).

III. Find a feasible T that yields the best performance.

Often in practice steps I–III are followed by step IV.

IV. If many satisfactory T exist or none exist, tighten or loosen the specifications accordingly and go to step I. Stop if this process cannot be carried further.

In this chapter we will emphasize steps II and III, especially when the performance function is of the "circular" type (for a definition see section 4.2). Step III requires the use of computer software, since numerical calculations are involved.[1] Step I was treated in Chapters 2 and 3, and step IV is illustrated in Chapter 5. See Chapter 5 for a discussion of performance functions that are not circular.

4.2 Generating a performance function

Let us suppose that we have one or more performance requirements that can be expressed as disk inequalities. We consider in this section the case where all of these can be put together as a single disk inequality, valid over all frequencies:

$$|K(j\omega) - T(j\omega)| \leq R(j\omega) \quad \forall \omega. \tag{4.1}$$

---

[1] The examples in this chapter were solved with the software package OTPDesign.

4.2. GENERATING A PERFORMANCE FUNCTION

We will build a "performance function" from (4.1) that can be used to solve design problems. Begin by rewriting (4.1) in the following way:

$$\frac{1}{R(j\omega)^2}|k(j\omega) - T(j\omega)|^2 \leq 1 \quad \forall \omega. \tag{4.2}$$

Now for a given T calculate the largest value of the left-hand side of (4.2):

$$\gamma(T) = \sup_\omega \frac{1}{R(j\omega)^2}|k(j\omega) - T(j\omega)|^2. \tag{4.3}$$

If a particular T is available, then one can check if the original performance requirements are satisfied. To do so one just calculates $\gamma(T)$. It is not hard to convince oneself that two possible outcomes of calculating $\gamma(T)$ are:

$$\begin{aligned} \gamma(T) \leq 1 &\Rightarrow T \text{ satisfies (4.1)} \\ \gamma(T) > 1 &\Rightarrow T \text{ does not satisfy (4.1)}. \end{aligned} \tag{4.4}$$

Example. Suppose that the center k and radius r are given by

$$\begin{aligned} k(\omega) &= -1/2 \frac{j\omega}{(j\omega-1)^2} \\ r(\omega) &= \frac{2}{(\omega^2+1)^{1/2}}. \end{aligned} \tag{4.5}$$

We now check if

$$T(j\omega) = \frac{1}{j\omega + 1}$$

satisfies the disk inequality (4.1):

$$\begin{aligned} g(\omega) &:= \tfrac{1}{r(j\omega)^2}|k(j\omega) - T(j\omega)|^2 \\ &= \sup_\omega \tfrac{(\omega^2+1)}{4}| - 1/2\tfrac{j\omega}{(j\omega-1)^2} - \tfrac{1}{(\omega^2+1)^{1/2}}|^2 \\ &= \tfrac{4+21\omega^2+9\omega^4}{16(1+\omega^2)^2}. \end{aligned} \tag{4.6}$$

After some calculations one obtains

$$\frac{dg}{d\omega} = \frac{13\omega - 3\omega^3}{8(1+\omega^2)^3}. \tag{4.7}$$

Thus $g(\omega)$ has three critical points: $\omega = 0$, $\omega = \pm\sqrt{13}/3$. The value $g(0) = 1/4$ is a minimum, and $g(\pm\sqrt{13}/3) = 297/512$ is a maximum. Also, $g(\omega) \to 9/16$ as $\omega \to \infty$. A plot is most informative (see Fig. 4.2). Thus

$$\gamma(T) = \sup_\omega g(\omega) = \frac{297}{512} < 1.$$

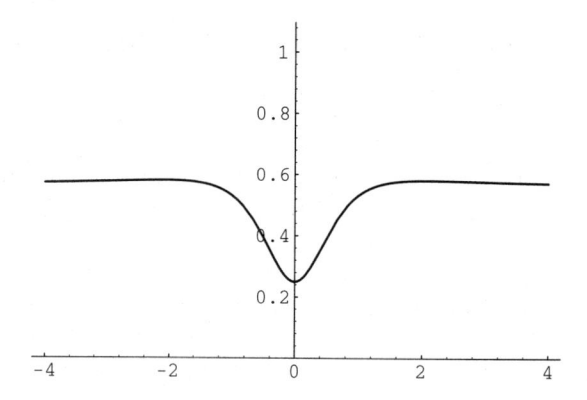

Fig. 4.2. *Plot of the function* $g(\omega) = 1/r(\omega)^2 |k(j\omega) - T(j\omega)|^2$.

We conclude that T does satisfy the disk inequality (4.1).

For designable T and given center and radius functions k and r, respectively, the expression

$$\frac{1}{R(j\omega)^2} |k(j\omega) - T(j\omega)|^2 \qquad (4.8)$$

is called the *circular performance function*, or simply *performance function*. The number $\gamma(T)$ is the *worst-case performance of* T, but most of the time we will simply call it the *performance of* T.

4.3 Finding T with best performance

A feasible function T^* with the property that

$$\gamma(T^*) \leq \gamma(T) \quad \forall T \text{ feasible} \qquad (4.9)$$

is called *optimal*. We reserve the symbol γ^* for the "optimal performance," that is, the performance of the optimal designable transfer function:

$$\gamma^* = \gamma(T^*). \qquad (4.10)$$

The optimal designable function T^* is a precious object since there is no other T that has better performance. By calculating the best performance possible, γ^*, the designer has conquered a peak. This vantage point offers information that is crucial in making decisions on the next steps. If T^* satisfies the constraints, one can use T^* as the choice of design or modify T to get a low-order model that still satisfies the constraints. If T^* does not satisfy the constraints, then is immediately known that there is no T that satisfies the constraints! The designer then can take other actions, such as to modify the "soft" constraints and then do the calculation again.

The process of finding the optimal T^* and the corresponding optimal performance γ^* is called *optimization*. The optimal T minimizes $\gamma(T)$ over all possible

4.3. FINDING T WITH BEST PERFORMANCE

T (recall that this includes internal stability). More precisely, we say that T^* is a solution to

$$\gamma^* = \inf_T \sup_\omega \frac{1}{R(j\omega)^2}|k(j\omega) - T(j\omega)|^2. \quad (4.11)$$

Because optimization is a complicated procedure, to carry it out one needs a computer with the appropriate software. When computer algorithms and software are not available to do the optimization, one has to resort to unsatisfactory approaches, such as writing a formula for T as a rational function with given order and unknown coefficients. Then one tries to do the get a good T by simply trying lots of possible choices of coefficients. Such methods are not always practical, and more important, the designer usually gets no indication of how far from optimal the choice of T is.

4.3.1 Example

Physically the problem is to stabilize the plant and impose closed-loop gain-phase margin and roll-off performance.

Problem. Solve *Design* for the plant $P(s) = (s+5)/(s+1)^2$, with performance requirements gain-phase margin $m = 0.5$ and closed-loop roll-off $|T(j\omega)| < |P(j\omega)|$, for $1 < \omega$.

Solution. Observe that the plant is strictly proper and stable. At this point we will not spend more time discussing internal stability, since the software we intend to use for the optimization takes care of this automatically.

We first construct the center and radius functions for our problem. Recall that the gain-phase margin constant m is the closest acceptable distance from the function $L = PC$ to the point -1; in other words,

$$|1 - T(j\omega)| < \frac{1}{0.5} = 2, \quad \text{for all } \omega.$$

The requirement on the roll-off can be written as

$$|T(j\omega)| < 0.707|P(j\omega)| \quad \text{for } 1 < \omega.$$

Thus the frequency domain performance requirements are described by (see Fig. 4.3)

$$\begin{array}{rll} |1 - T(j\omega)| & < \ 2 & \text{for } 0 \leq \omega \leq 1, \\ |T(j\omega)| & < \ 0.707|P(j\omega)| & \text{for } 1 < \omega. \end{array} \quad (4.12)$$

The inequalities in (4.12) can be cast as a single disk inequality

$$|K(j\omega) - T(j\omega)| < R(j\omega),$$

where

$$K(j\omega) = \begin{cases} 1 & \text{if } |\omega| < 1 \\ 0 & \text{if } |\omega| \geq 1 \end{cases} \quad (4.13)$$

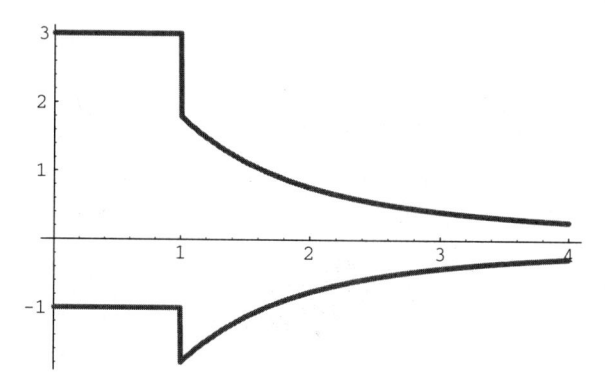

Fig. 4.3. *The performance envelope defined by center and radius functions in the example is outlined by the thick curves. The horizontal axis is frequency ω, and the two functions are $K(j\omega) + R(j\omega)$ and $K(j\omega) - R(j\omega)$.*

and
$$R(j\omega) = \begin{cases} 2.0 & \text{if } 0 \le |\omega| < 1 \\ 0.707|P(j\omega)| & \text{if } 1 < |\omega|. \end{cases} \quad (4.14)$$

The optimal T^* is the solution to

$$\gamma^* = \inf_{T \in \mathcal{I}} \sup_\omega \frac{1}{R(j\omega)^2} |K(j\omega) - T(j\omega)|^2. \quad (4.15)$$

To give an idea of what the answer is, some results of calculations with the software package OPTDesign are presented below. For more details see Chapter 5.

Calculations by computer give $\gamma^* = 0.09923$, so there exist solutions to *Design*. Here is a rational function that is a low-order approximation to the optimal one:

$$T_0(s) = \frac{5-s}{(s+1)^2} \frac{(0.0311 + 0.2613s)}{(0.8809 + s)}. \quad (4.16)$$

See Figs. 4.4–4.7 for plots of the optimal T^*.

4.4 Acceptable performance functions

There are two basic rules that have to be followed to define a useful performance function.

RULE 1: The performance at any T and any frequency ω (including $\omega = \infty$) should be a well-defined real number.[2]

---

[2] The performance may turn out to be infinitely large at frequencies ω_0 such that the plant has a zero at $s = j\omega_0$, or at infinite frequency. This should not be a problem, provided that only T that give internally stable systems are considered.

4.4. ACCEPTABLE PERFORMANCE FUNCTIONS

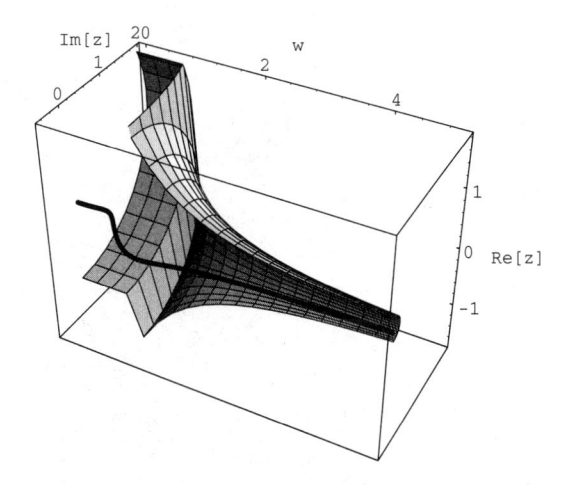

Fig. 4.4. *3-D plot of the performance envelope and the solution T_0. Clearly T_0 satisfies the constraints.*

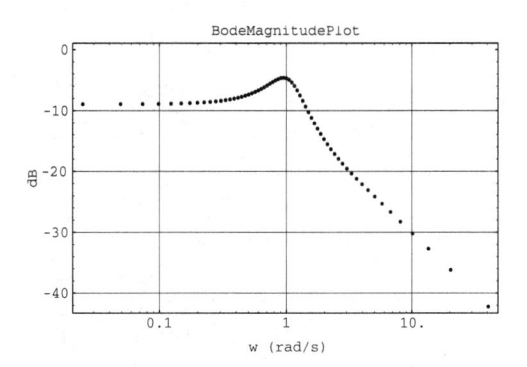

Fig. 4.5. *Magnitude of T_0.*

RULE 2: For any fixed frequency ω, the performance as a function of T cannot be independent of T.

The justification of rules 1 and 2 requires the results in Part III of this book, so we refer the interested reader to it.

The following are examples of performance functions that are common in the engineering literature. We assume that the underlying design problem has internal stability requirements, and that the plant P is strictly proper and has no poles or zeros on the $j\omega$ axis. Also, W, W_1, and W_2 denote given rational weight functions of frequency.

1. $\Gamma(\omega, T(j\omega)) = W(j\omega)|T(j\omega)|$. If the system has a plant P with relative degree p, then W must have relative degree $-p$ (or larger) for the per-

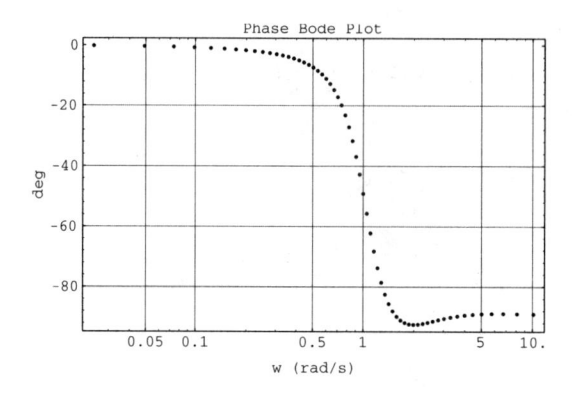

Fig. 4.6. *Phase of T_0.*

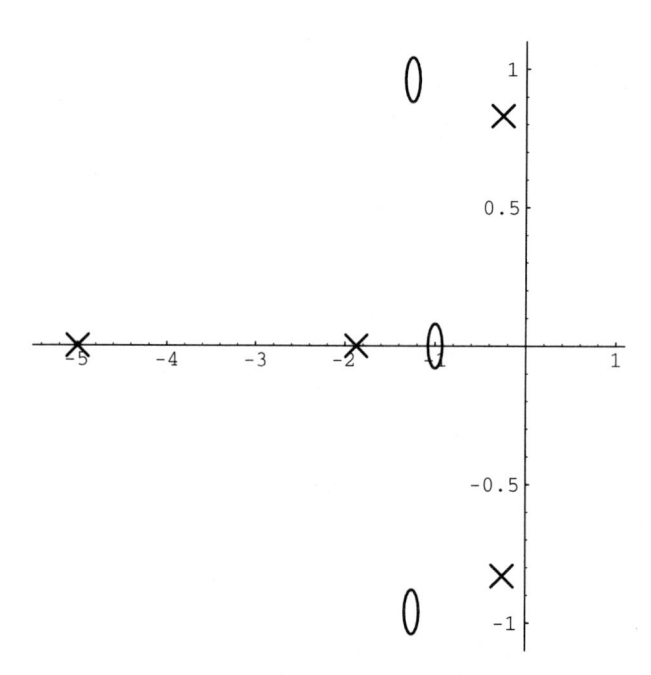

Fig. 4.7. *The zeros and poles of the compensator $(0.786826 + 2.36403s + 2.68265s^2 + 1.42052s^3 + 0.315071s^4)/(7.08474 + 9.97784s + 13.6385s^2 + 7.38526s^3 + s^4)$ that corresponds to T_0 are indicated by "o" and "x," respectively. Note that this compensator is stable. Stability of the compensator is not guaranteed a priori by the design method of this book.*

4.5. PERFORMANCE NOT OF THE CIRCULAR TYPE

formance to be well defined. If W does not, then Rule 2 is violated at $\omega = \infty$.

2. $\Gamma(\omega, T(j\omega)) = W(j\omega)|1 - T(j\omega)|$. This is not a good performance, no matter what W is. The reason is that since T rolls off at high frequency, the performance at very high frequency is determined by W only. This performance has no way of measuring how fast T rolls off or any relevant high-frequency property of T.

3. $\Gamma(\omega, T(j\omega)) = W_1(j\omega)|T(j\omega)| + W_2(j\omega)|1 - T(j\omega)|$. The weight W_2 should be 0 at infinite frequency, so that the term with this weight does not influence performance at very high frequency. The comment in the first example also applies to the weight W_1.

4. $\Gamma(\omega, T(j\omega)) = W_1(j\omega)|T(j\omega)|^2 + W_2(j\omega)|1 - T(j\omega)|^2$. This continues example 3. The relative degree of W_2 should be $-p/2$ or larger, where p is the relative degree of the plant.[3]

4.5 Performance not of the circular type

In this section we discuss performance functions more general than (4.8). Suppose that we have a design problem where one of the constraints is

$$W_1(\omega)|T(j\omega)| + W_2(\omega)|1 - T(j\omega)| < 1, \quad \omega_1 \leq \omega \leq \omega_2, \quad (4.17)$$

where W_1 and W_2 are known weight functions.[4] Clearly this constraint is not a disk inequality. Indeed, the region

$$S_\omega(c) = \{z : W_1(\omega)|z| + W_2(\omega)|1 - z| \leq c\}$$

is not a disk. However, the inequality (4.17) can be used to define a performance function valid at least for those ω satisfying $\omega_1 \leq \omega \leq \omega_2$. For this we set, for z any complex number,

$$\Gamma(\omega, z) = W_1(\omega)|z| + W_2(\omega)|1 - z|.$$

Thus a designable transfer function T satisfies the constraint at frequency ω if and only if

$$\Gamma(\omega, T(j\omega)) \leq 1. \quad (4.18)$$

The function Γ is called a *performance function*. Let us define for $T \in \mathcal{RH}^\infty$ the number

$$\gamma(T) := \sup_\omega \Gamma(\omega, T(j\omega)). \quad (4.19)$$

---

[3] When zeros or poles of the plant occur on the $j\omega$ axis, weight functions have to be chosen so that the weights in the terms $W_1(j\omega)|T(j\omega)|^2$ and $W_2(j\omega)|1-T(j\omega)|^2$ have the zeros (resp., poles) at the same values of ω as those of the plant, order included. The reason for this is the internal stability requirement. See Chapter 7

[4] See [OZ93]. Also see section 6.5 in Chapter 6.

The number $\gamma(T)$ represents the cost, or overall performance, associated with the designable transfer function T. It is called the *performance index*. Note that (4.18) can be written as

$$\gamma(T) \leq 1. \tag{4.20}$$

Observe that $\gamma(T) < 1$ means that T satisfies the performance requirements, with some slack. Thus we associate small values of $\gamma(T)$ with *good performance*.

The simplest functions Γ arise when a single disk inequality applies at each frequency.

4.6 Optimization

It is a fact that there is a formula that depends on the RHP poles and zeros of the given plant P that gives all $T \in \mathcal{I}$. Indeed, there exist functions $A, B \in RH^\infty$ such that

$$T \in \mathcal{I} \iff T = A + BT_1 \quad \text{for some } T_1 \in RH^\infty. \tag{4.21}$$

For general P this formula is proved in Chapter 7, while for RHP-stable P we already saw in Chapter 2 that

$$T \in \mathcal{I} \iff T = PT_1 \quad \text{for some } T_1 \in RH^\infty. \tag{4.22}$$

One approach to treating the design problem in practice is to use formulas (4.21) and (4.22) to account for internal stability. Depending on the specific approach and the tools available, to solve *Design* the formulas are either manipulated directly by the designer or they are automatically handled by the computer. In this case only the list of unstable zeros and poles of the plant is necessary for the software to build and use the necessary formulas.

4.6.1 The optimization problem $OPT_\mathcal{I}$

Suppose that steps I and II in Section 4.1 have been completed; that is, there is available a performance function Γ and a set $\mathcal{I}$ of admissible functions. To find the "best" T possible, one must find the smallest possible value for the performance index $\gamma(T)$ for all $T \in \mathcal{I}$. That is, one must solve the following problem:

$OPT_\mathcal{I}$ Given $\Gamma(\omega, z)$, a positive-valued function of $\omega \in \mathbb{R}$ and of $z \in \mathbb{C}$, and a set $\mathcal{I}$ of admissible functions, find

$$\gamma^* := \inf_{T \in \mathcal{I}} \sup_\omega \Gamma(\omega, T(j\omega)) \tag{4.23}$$

and find an optimal T^* if it exists.

In general $OPT_\mathcal{I}$ is a constrained problem, since one may have $\mathcal{I} \neq RH^\infty$. By solving $OPT_\mathcal{I}$ we solve *Design*. This follows from Proposition 4.6.1.

4.7. INTERNAL STABILITY AND OPTIMIZATION

PROPOSITION 4.6.1. *If $\gamma^* < 1$, then there exist solutions to Design, and if $\gamma^* > 1$, then there are no solutions to Design.*

Proof. By definition, the number γ^* is the smallest value possible for γ so that the inequality
$$\Gamma(\omega, T(j\omega)) \leq \gamma \tag{4.24}$$
has solutions $T \in \mathcal{I}$. Therefore $\gamma^* > 1$ says that there are no functions Γ that satisfy (4.20), i.e., that there are no solutions to *Design*. Similarly, $\gamma^* < 1$ says that there is at least one function $T \in \mathcal{I}$ that satisfies (4.20), which implies that there exist solutions to *Design*. ∎

4.7 Internal stability and optimization

It is a very simple matter to transform $OPT_\mathcal{I}$ into an unconstrained optimization problem OPT. This is desirable because powerful mathematical algorithms for solving unconstrained problems are available. It is so easy that the computer can do this in a manner that is transparent to the user, for example, as implemented in the package OPTDesign. Occasionally the designer may want to do it directly by hand (or may be forced to do it). Thus we include a discussion of this below.

4.7.1 The optimization problem OPT

To do the transformation of $OPT_\mathcal{I}$ into OPT we use the formula (4.21) in combination with the performance $\Gamma(\omega, z)$. Suppose T is as in (4.21). Define a new performance function as follows:

$$\Gamma_1(\omega, z_1) = \Gamma(\omega, A(j\omega) + B(j\omega) z_1) = \Gamma(\omega, z). \tag{4.25}$$

Therefore $OPT_\mathcal{I}$ is equivalent to the following unconstrained optimization problem:

OPT Given $\Gamma_1(\omega, z)$, a positive-valued function of $\omega \in R$ and of $z \in C$, find

$$\gamma_1^* := \inf_{T_1 \in \mathcal{I}} \sup_\omega \Gamma_1(\omega, T_1(j\omega)) \tag{4.26}$$

and find an optimal T_1^* if it exists.

4.7.2 OPT with circular Γ

Note that if the function Γ is circular, then the function Γ_1 in (4.25) is also circular. To see this, combine the disk inequality (4.2) with relation (4.21) to obtain the inequality

$$\frac{1}{R(j\omega)^2} |K(j\omega) - A(j\omega) - B(j\omega) T_1(j\omega)|^2 \leq 1 \quad (\forall \omega). \tag{4.27}$$

The resulting inequality is itself a disk inequality in the variable T_1:

$$\frac{|B(j\omega)|^2}{R(j\omega)^2}\left|\frac{K(j\omega)-A(j\omega)}{B(j\omega)}-T_1(j\omega)\right|^2 \leq 1 \quad (\forall \omega). \tag{4.28}$$

Thus we can choose $\Gamma_1(\omega, T_1(j\omega))$ to be the left-hand side of inequality (4.27) to arrive at the unconstrained optimization problem OPT.

4.8 Exercises

1. Is the Γ given below of quasi-circular type? Take $k(j\omega) = \frac{j\omega+2}{j\omega-3}$.

    (a) $\Gamma(j\omega, z) = |z-3|^2 + |z-(2+j)|^2 + |z-[2-j]|^2$
    (b) $\Gamma(j\omega, z) = 4\operatorname{Re}(z-k(j\omega)) + 16\operatorname{Im}(z-k(j\omega))$
    (c) $\Gamma(j\omega, z) = |0.8 + (z-k(j\omega))^2|^2$
    (d) $\Gamma(j\omega, z) = 3\left|\frac{z-k(j\omega)}{z+v(j\omega)}-6\right|^2 + \left|\frac{z-k(j\omega)}{z+v(j\omega)}+2\right|^2$, where $v(j\omega) = \frac{2j\omega+1}{5j\omega-2}$

2. Is $\Gamma(j\omega, z) = \frac{|z-k(j\omega)|^2}{r(j\omega)^2}$ with k and r given below nondegenerate?

    (a) $r(j\omega) = \frac{1}{2}$

    $$k(j\omega) = \begin{cases} 1 & |\omega| \leq 3 \\ 2 & 3 < |\omega| \leq 6 \end{cases}$$

    (b) $r(j\omega) = \frac{|(j\omega-i)|^2}{|j\omega+1|^2}$

    $$k(j\omega) = \begin{cases} 1 & |\omega| \leq 3 \\ 0 & |\omega| > 3 \end{cases}$$

ANSWERS

1a Yes

b No, the level circular are ellipses.

c No, the level curves are lemniscates.

d Yes, because linear function transforms take circles to circles.

2. All Γ are degenerate, because

a Γ is not defined at all frequencies. This is very bad.

b $\Gamma(j) = \infty$. This causes numerical problems. You must change variables in function space (see Part III of this book).

Chapter 5

A Design Example with OPTDesign

5.1 Introduction

This chapter introduces the reader to practical design by showing how the ideas learned so far feed into a computer implementation. It is written in a generic tone, so that one need not know anything specific about computation to get a concrete idea of how to do a control design. The best explanation of this subject is to give an actual computer design session. Unfortunately, this usually entails getting involved in many specialized details peculiar to a package. However, we have developed a program whose use is close enough to conventional English that anyone can read it without specialized knowledge.

Our program is called OPTDesign and runs under Mathematica. This permits any standard symbolic or numerical calculations to be done in a language that is quite compatible with standard mathematical notation.

5.2 The problem

We now give the data for a specific problem. The plant is

$$P(s) = \frac{(1 - s/5)}{(1 + s/2)^2}$$

and the requirements are tracking on the band $0 \leq \omega \leq 0.3$ with bound 1.5, gain-phase margin constant of 0.5, bandwidth $0 < \omega \leq 2.0$; and closed-loop roll-off with bound 2.5 on the band $4.0 \leq \omega$ (see Table 5.1). We wish to design an internally stable system $\mathcal{S}$ that satisfies the given performance requirements.

The following is a list of inputs for a computer run. If you wanted to try OPTDesign you would begin by loading OPTDesign into a Mathematica session.

```
<<OPTDesign`
```

Table 5.1. *Performance requirements.*

Constraint				Band
Tracking	$\lvert T(j\omega) - 1\rvert$	<	α_t	$0 \leq \omega < \omega_t$
Gain-phase	$\lvert T(j\omega) - 1\rvert$	<	α_{gpm}	$\omega_t \leq \omega < \omega_b$
Bandwidth	$\lvert T(j\omega)\rvert$	<	α_b	$\omega_b \leq \omega < \omega_r$
Roll-off	$\lvert T(j\omega)\rvert$	<	$\alpha_r \lvert P(j\omega)\rvert$	$\omega_r \leq \omega < \infty$

We enter the center and radius functions k_0 and r_0 directly as step functions. For example, in Mathematica this is done using the *Which[ ]* command. This is a strange name for what most English speakers call "If."

```
      p[s_] = ( 1 - s/5 )/( 1 + s/2 )^2;

       wt = 0.3;
    alphat = 1.5;
  alphagpm = 2.0;
       wb = 2.0;
    alphab = 0.7;
       wr = 4.0;
    alphar = 2.5;

  k0[w_] = Which[ 0  <=  Abs[w] < wb, 1.0,
                 wb  <=  Abs[w]     , 0.0];

  r0[w_] = Which[0   <=  Abs[w] < wt, alphat,
                 wt  <=  Abs[w] < wb, alphagpm,
                 wb  <=  Abs[w] < wr, alphab,
                 wr  <=  Abs[w]     , alphar Abs[ p[I w] ] ];
```

The requirements envelope that we defined above has jump discontinuities at two locations on the semiaxis $\omega > 0$. For sampling functions, a grid of points on the $j\omega$ axis has to be chosen. If the user does not specify a grid, then OPTDesign chooses one with 128 points distributed around $\omega = 1$.[1] Also, OPTDesign automatically does a very small amount of smoothing of the discretized center and radius functions (if desired, the user may specify the amount of smoothing). The plot of the requirements envelope before and after smoothing gives an idea

---

[1] To change the default grid on the $j\omega$ axis, use `SetGrid[n,GridSpread -> b]`, where n is an integer representing the number of gridpoints, and b is a positive number representing the frequency around which the gridpoints are distributed.

5.2. THE PROBLEM

of how much distortion we are introducing into the problem by smoothing functions. We must keep in mind that the solution obtained with OPTDesign will be optimal with respect to the "smoothed" envelope. The following command produces the plot shown in Fig. 5.1.

```
EnvelopePlot[Radius->r0,Center->k0, FrequencyBand->{0.,wr+1.}]
```

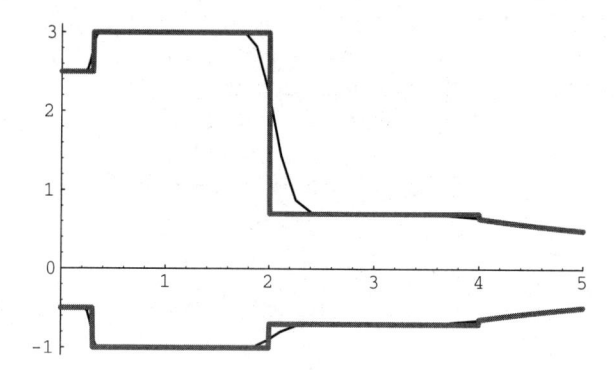

Fig. 5.1. *The original and smoothed requirements envelopes.*

A 3-D plot of the requirements envelope (see Fig. 5.2) is produced with

```
EnvelopePlot3D[Radius->r0,Center->k0]
```

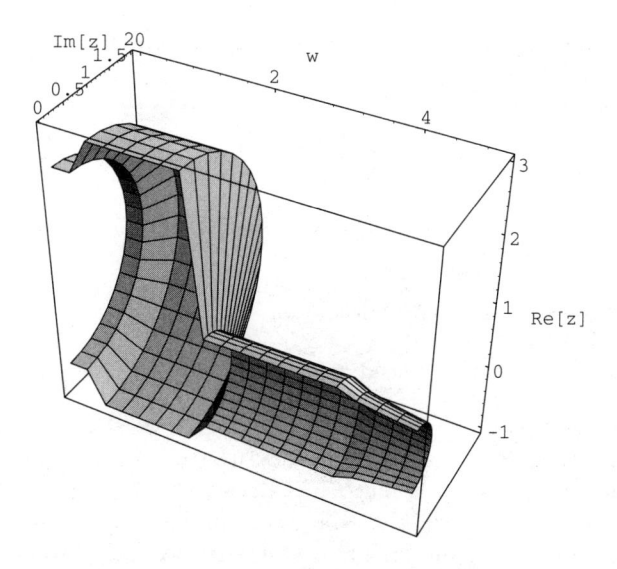

Fig. 5.2. *3-D plot of the requirements envelope.*

A plot of the discrete profile of the requirements envelope can be used to judge if there are enough frequency gridpoints in the band of interest (see Fig.

5.3). Note that in our example most of the gridpoints are in the band $0.1 < \omega < 10.0$, which is where the center and radius functions have their distinctive features. This is desirable.

```
EnvelopeLogPlot[Radius->r0,Center->k0,
        FrequencyBand->{0.,wr+1.},Discrete -> True];
```

If we do not like either of the plots in Figs. 5.1 and 5.3, then several input parameters can be modified to rerun the problem and to obtain more satisfactory plots. We leave this for a later section and proceed now with a run of the program OPTDesign with the given data.

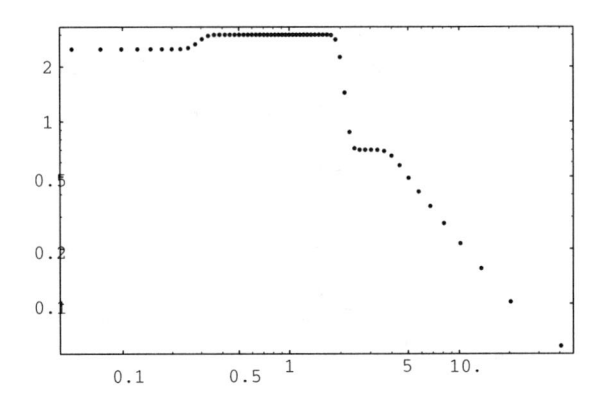

Fig. 5.3. Plot of the discrete profile of the requirements envelope.

5.3 Optimization with OPTDesign

Running OPTDesign for plant P and specified disks with radius r_0 and center k_0 is done by entering

```
OPTDesign[p,Radius->r0,Center->k0]
```

The main things that are produced are

- A parameter γ^* that tells us if (a smoothed version of) this disk-type problem has a solution. If γ^* is less than 1, then the problem has a solution; indeed if γ^* is much less than 1, then the constraints can be tightened. But if $\gamma^* > 1$, then the design problem has no solution and the constraints must be loosened to obtain a problem that has a solution.

- The *optimal closed-loop transfer function T*.

Diagnostics and progress reports are routinely printed to the screen as the program OPTDesign runs:

5.3. OPTIMIZATION WITH OPTDESIGN 51

```
Parameterization for internal stability:

T = A + B * T1, where T1 is RHP stable and

         -5 + s
A[s] =---------
             2
        (2. + s)


         -5 + s
B[s] =---------
             2
        (2. + s)




Processing performance function...
Sampling Radius and Center ...
Smoothing Radius and Center ...
Optimization routine Anopt now number-crunching ...

It :   Current Value   :  Step   : Optimality Tests :  Error  : Sm. : Grid
   :      gamma        :        :   Flat    GrAlign :   ned   :     :
------------------------------------------------------------------------------
 0 : 2.2485924730083E+00 :  N/A   : 8.9E-01 : 1. E+00 :   N/A   : NON : 128
 1 : 1.5969785364195E-01 : 1.8E+00 : 3.6E-03 : 0  E+00 : 1.5E-02 : NON : 128



        Summary
        -------
        Supremum of gamma: gamma*  = 1.596978536419496E-01
        Optimality Test  : Flat    = 3.5999393585E-03
        Optimality Test  : GrAlign = 0
        Error diagnostic : ned     = 1.54E-02

        Calculating output functions ...
        Resetting options ...
        Done!
```

Observe that the output parameter γ^* (in the above output, *gamma**) is less than 1. The column *ned* in the screen output is a measure of numerical error or noise. *Flat* and *GrAlign* are diagnostics. If they are near 0, the calculated solution is nearly optimal. We say more on this later in section 5.6.

We conclude that there are solutions to the (smoothed) *Design* problem and that the accuracy of the computation is acceptable. Furthermore, since γ^* is much less than 1, we can tighten the performance requirements substantially and still get a solution. The calculated closed loop transfer function, open loop, and compensator output is stored in the output variables T, L and Co respectively. We postpone discussing their format until subsection 5.5.1, where we show how to plot and manipulate it. Now we move on to the production of a compensator.

5.4 Producing a rational compensator

The next step is to generate a rational compensator. We will first produce a rational expression for the closed-loop transfer function and extract a rational compensator from it.

Computer routines for doing rational approximation are not extremely reliable when the degree of the approximation is high (see section 5.9). The rational approximation in our package appropriate here is RationalModel. By default RationalModel finds a rational function that approximates the closed-loop transfer function.[2]

```
Trat[s_] = RationalModel[s , DegreeOfDenominator->3]

Error = 0.171731

  -5 + s       (-4.67853 - 1.92117 s) (-5 + s)
  --------  +  -------------------------------
        2              2
  (2. + s)      (2. + s)  (3.769 + 1. s)
```

RationalModel produces a rational function of the form

$$T_{rat}(s) = A(s) + B(s)T_{1rat}(s)$$

with stable function $T_{1rat}(s)$. The proper rational functions A and B depend on the plant and incorporate automatically the internal stability requirements on T. **You Must run OPTDesign before you run RationalModel.**

Recall that the compensator C and the closed-loop transfer functions are related by the formula

$$C = \frac{1}{P} \cdot \frac{T}{1-T}.$$

Here we call the compensator *comp1*.

```
comp1[s_] = 1/p[s] Trat[s]/(1-Trat[s]);
```

There is no point in writing out the formula for the output now, since most valuable to us will be a zero- and- pole plot of *comp1*. The following command produces the plot shown in Fig. 5.4.

```
PlotZP[comp1[s],s]
```

The figure suggests that *comp1* has a pole-zero pair at $s = 5$.[3] The standard Mathematica functions do not cancel terms with decimal notation. To do the cancellation we can use a function provided with OPTDesign:

---

[2] The *Error* number displayed by the RationalModel routine refers to the fit of T_1 and not to the fit of T.

[3] One way to confirm this is
```
comp1zeros = s /. NSolve[ Numerator[comp1]==0,s]
{-2., -2., -0.987364, 5. }
comp1poles = s /. NSolve[ Denominator[comp1]==0,s]
{5, -6.60224, -1.04396 - 0.710493 I, -1.04396 + 0.710493 I}
```

5.4. PRODUCING A RATIONAL COMPENSATOR

```
comp2[s_] = CancelZP[comp1,s - 5,s]
```

$$\frac{4.54766 + 9.15353\, s + 5.74278\, s^2 + 1.15147\, s^3}{10.5283 + 15.3797\, s + 8.69017\, s^2 + 1.\, s^3}$$

Hence *comp2(s)* results from the simplification of *comp1*. We plot the poles and zeros of *comp2* now (see Fig. 5.5).

```
PlotZP[comp2[s],s]
```

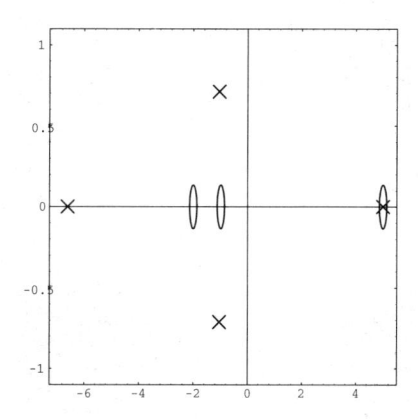

Fig. 5.4. *Poles and zeros of comp1.*

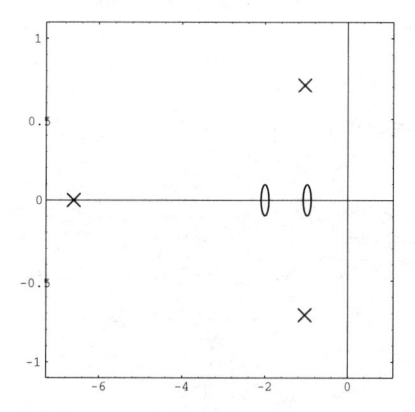

Fig. 5.5. *Poles and zeros of comp2.*

It is clear from the plots in Fig. 5.4 and Fig. 5.5 that the compensator *comp2* does not cancel RHP poles or zeros of the plant, which is a necessary condition for internal stability of the overall system $\mathcal{S}$. Now we calculate the closed-loop transfer function *Trat2* that corresponds to the compensator *comp2*, and follow this with a plot of the poles and zeros of *Trat2* (see Fig. 5.6).

```
Trat2[s_] = p[s] comp2[s] /(1+p[s] comp2[s]);

PlotZP[Trat2[s],s]
```

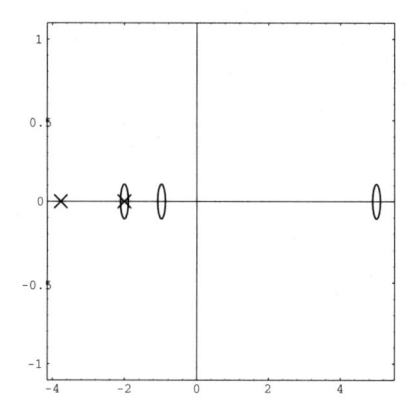

Fig. 5.6. *Poles and zeros of Trat2.*

Note that since *comp2* is proper and *Trat2* is stable, for this choice of compensator the system is internally stable.

5.5 How good is the answer?

To verify that we have met the performance requirements we plot the closed-loop transfer function and the performance envelope simultaneously. A 3-D picture (shown in Fig. 5.7) of the requirements envelope together with the plot of *Trat2* is produced with the following command.

```
EnvelopePlot3D[Radius->r0,Center->k0,ClosedLoop->Trat2]
```

For those who like old-fashioned 2-D plots, the following command produces the plot in Fig. 5.8.

```
Plot[ 1/r0[w] Abs[k0[w] - Trat2[I w]],{w,0,6}]
```

One usually wants to look at Bode plots to evaluate the design. These can be obtained in several ways in most control packages. For example, in OPTDesign the following commands produce the plots.

```
BodeMagnitude[ Trat2[s],s,{w,0.1,10},PlotLabel->"Magnitude Plot"];
BodePhase[ Trat2[s],s,{w,0.1,10},PlotLabel->"Phase Plot"];
```

5.5. HOW GOOD IS THE ANSWER?

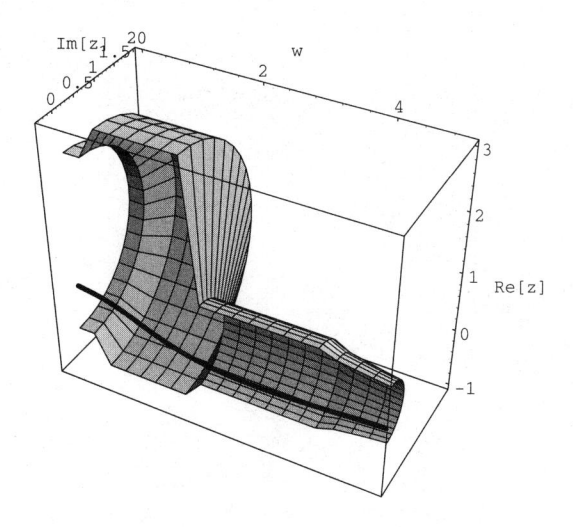

Fig. 5.7. *3-D plot of Trat2 inside the requirements envelope.*

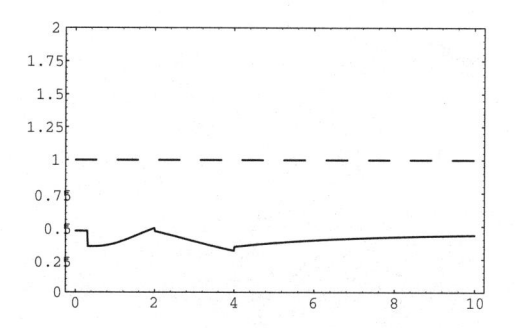

Fig. 5.8. *Plot of the original, (nonsmoothed) performance of Trat2.*

Instead of showing these we shall illustrate an important numerical point by demonstrating Bode plots of functions on the $j\omega$ axis gridpoints used by the latest run of OPTdesign. This allows one to see if the grid captures the frequency bands that are important for the problem. The Bode plots are shown in Figs. 5.9 and 5.10. The commands are as follows.

```
Tw = Discretize[Trat2];

BodeMagnitude[Tw,s,{w,0.1,10},PlotLabel->"Magnitude Plot"];

BodePhase[ Tw,s,{w,0.1,10},PlotLabel->"Phase Plot"];
```

Plots for the open loop $L = PC$ can be produced in similar fashion by plotting $L = pcomp2$.

One can see in the magnitude plot of the closed loop that most of the sample points are located in the frequency band $0.1 < \omega < 10$. One can judge this to

```
              Magnitude Plot
      ┌─────┬──────┬──────┬──────┐
  -10 │     │   ....··········...│
      │     │..···         ·     │
  -11 │·····│                ·   │
      │     │                    │
Bd -12├─────┼──────┼──────┼──────┤
      │     │                ·   │
  -13 │     │                    │
      │     │                 ·  │
  -14 ├─────┼──────┼──────┼──────┤
      │     │                    │
  -15 │     │                 ·  │
      └─────┴──────┴──────┴──────┘
         0.1     1      10.

              w (rad/s)
```

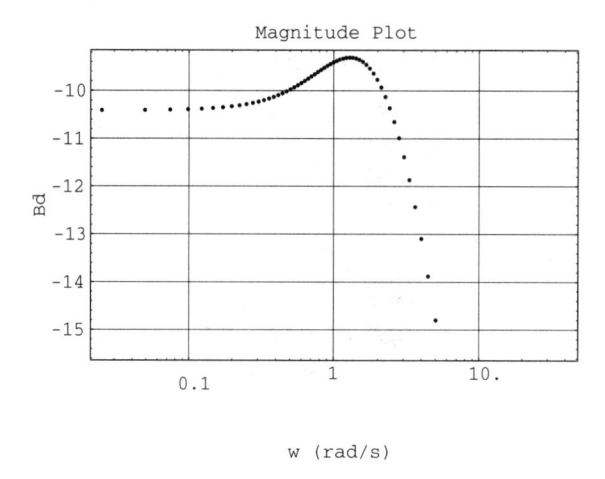

Fig. 5.9. *Bode plot (magnitude) of the closed-loop transfer function Trat2.*

be acceptable from the numerical standpoint, based on the fact that both the input functions k_0 and r_0 change mostly in this band. For details see section 5.8.

5.5.1 More on plots and functions

Some users may want to manipulate the output of an OPTDesign run before dealing with rational fits. If you type T in a session after you run OPTDesign, then Mathematica returns a list of complex numbers that are the values of T on the grid. Now we give a brief discussion of commands to plot and manipulate the closed loop T, the open loop L, or other lists of data. A list of some useful commands is *EnvelopePlot3D, BodeMagnitude, BodePhase, Nyquist, Nichols, RHPListPlot3D, OPTDParametrize, Discretize, Grid*.

To plot the list of data T and the requirements envelope simultaneously, type

```
EnvelopePlot3D[ Radius -> r0, Center -> k0, ClosedLoop -> T]
```

The important point is that the call is the same as whether T is a rational function or a list of data. The commands

```
BodeMagnitude[T], BodePhase[T], Nyquist[L], Nichols[L]
```

also have this feature of working either on functions defined by formulas or lists of data.

If one wants to plot or manipulate T by itself, then one must access the grid on the $j\omega$ axis where your OPTDesign session is evaluating T, via

```
Grid[ ]
```

5.5. HOW GOOD IS THE ANSWER?

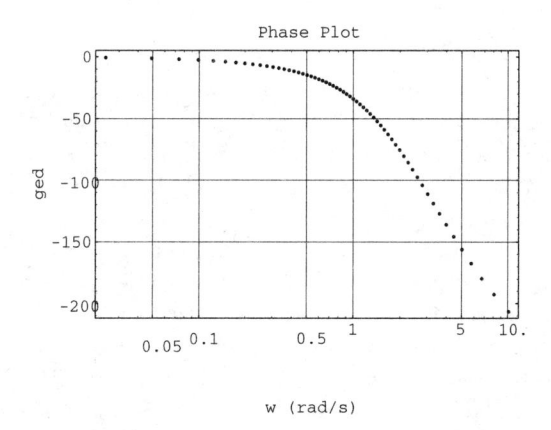

Fig. 5.10. *Bode plot (phase) of the closed-loop transfer function Trat2.*

The output is a list of numbers on the $j\omega$ axis. Now we can plot T using

```
RHPListPlot3D[ T, PlotRange -> {{0,3},Automatic,Automatic}]
```

where the PlotRange option sets the range you will see.

One can put together a functions values with the grid on which it is defined, using

```
OPTDParametrize[T]
```

Note that you may achieve identical results with

```
Transpose[{Grid[],T}]
```

Indeed, this is how *OPTDParametrize* is defined.

It might well be useful to note that the plotting commands above have a very simple core plus a few embellishments to make the scales, to make labels, to put the -1 point into the Nyquist plot, etc. As an example we illustrate how one builds plotting routines such as *Nyquist* and *RHPListPlot3D*. The core of `RHPListPlot3D[ T ]` is

```
ScatterPlot3D[ Transpose[ {Grid[], Re[T], Im[T]} ] ]
```

The core of `Nyquist[ L ]` is

```
ListPlot[ Transpose[ {Re[L], Im[L]} ]]
```

So far we have discussed functions presented as data sets. Now we mention that if you have a rational function T_r rather than a lot of discrete values, `Discretize[$T_r$]` gives a list of values of T_r on the ambient OPTDesign session grid.

The commands mentioned in this section and their output are discussed further in Appendix G.

5.6 Optimality diagnostics

If a computer run stops, does it mean that a solution has been obtained? How good is the "calculated solution"? If garbage was produced, how do we know? These questions can be answered with *optimality diagnostics*. These are one or more indicators that tell directly or indirectly whether a "computed solution" is close to an "actual solution."

The importance of having optimality diagnostics available for optimization problems cannot be overemphasized. Any respectable optimization theory must have them, and any serious optimization software must implement them.

The optimization theory behind the problem *Design* has two diagnostics that are crucial: the *Flatness diagnostic* and the *Gradient Alignment diagnostic*. These are positive real numbers; when the design is *a true theoretical optimum*, then these two numbers are equal to zero. They could be implemented in most H^∞ packages we know, so they are in no way dependent on OPTDesign.

When run, the program OPTDesign "iterates," that is, produces a sequence $T^{(0)}, T^{(1)}, T^{(2)}, \ldots$, of "guesses at the answer," which improve the performance with each step or iteration. For each one of these iterations, the Flatness and Gradient Alignment diagnostics are printed on the screen so that the user may follow the progress of the computer run.

The Flatness and Gradient Alignment diagnostics are labeled *Flat* and *GrAlign*. Other output diagnostics that are printed on the screen for each iteration are the performance level attained at such iteration, γ, and an indicator of numerical error, *ned*. Table 5.2 gives more details on this.

5.7 Specifying compensator roll-off

In the example above we obtained a compensator with relative degree 0; that is, the compensator was proper but not strictly proper, so it does not roll off. Recall from section 3.2.4 that if you want to specify the relative degree of the compensator to be $d(C) \geq 1$, set the roll-off constraint so that the radius function goes to 0 asymptotically with

$$\frac{1}{\omega^{[d(C)+d(P)]}}$$

as $\omega \to \infty$, where $d(P)$ is the relative degree of the plant P.

Here is an example where the relative degree of the compensator is specified as $d(C) = 2$. Note that the roll-off of the function $|p(j\omega)/\omega^2|$ equals $d(P) + 2$, which is the appropriate roll-off of the radius. Observe that the degree of the compensator enters in two places, in the definition of the radius function and in the call to OPTDesign.

```
k0[w_] = Which[0   <= Abs[w]  < wb, 1.0,
               wb <= Abs[w]      , 0.0];
```

5.8. REDUCING THE NUMERICAL ERROR

```
r0[w_] = Which[0  <= Abs[w] < wt, alphat,
               wt <= Abs[w] < wb, alphagpm,
               wb <= Abs[w] < wr, alphab,
               wr <= Abs[w]     , alphar Abs[p[I w]/w^2]
              ];

OPTDesign[p,Center->k0,Radius->r0,Cdecay -> 2];
```

Table 5.2. *Optimality diagnostics.*

Diagnostic	Meaning
γ	A positive number. It is the performance level attained at a current guess T at the answer. More precisely, it gives the worst-case performance value. If $\Gamma(\omega, z)$ is the performance function and $T(j\omega)$ is the current guess at the answer, then $\gamma = \gamma(T) = \sup_\omega \Gamma(\omega, T(j\omega))$.
Flat	A number between 0 and 1. One of the fundamental results in the theory of H^∞ optimization is the fact that in most problems, an optimal design flattens the performance (see Chapter 9 in [HMer98]). That is, if T^* is an optimal function, a plot of $\Gamma(\omega, T^*(j\omega))$ vs. ω produces a horizontal line. The diagnostic flat measures the "nonflatness" of the current guess at the answer. To give a hypothetical example, if the performance at the current guess varies between 0.25 and 0.75 as ω goes from 0 to ∞, then in this case the flatness diagnostic is defined as $Flat = \frac{0.75 - 0.25}{0.75} = 0.375$ One interpretation is that the range of the performance at the current guess is 37.5% of its maximum value.
GrAlign	A nonnegative number. This optimality diagnostic at the true answer is equal to 0. It verifies that a certain winding number has the "appropriate value." For details see sections 9.5 and 13.5 in [HMer98].
ned	A positive number. When small it indicates that numerical calculations contain little numerical error. Numerical trouble and what to do about it are discussed in section 5.8.

5.8 Reducing the numerical error

In a nutshell, numerical error might be too large as a result of one or more of the following causes: the requirements envelope changes too fast (discontinuities is a reason that comes to mind), the grid on the $j\omega$ axis is badly chosen, or the rational approximation to the solution to the problem is poor. In this section we give some pointers as to how to proceed if numerical error arises.

1. *The choice of gridpoints on the $j\omega$ axis.* The command `Grid[Ngrid` $\to$ n, `GridSpread` $\to b$ `]` produces n gridpoints on the ω axis, which ac-

cumulate near $\omega = b$. For OPTDesign to perform well when solving a design problem, it is necessary that a "sufficient" number of gridpoints be produced in a frequency band where the center and radius functions have their distinctive features. If this is a somewhat large band, then one chooses b to be a midpoint of this band and also chooses n large enough to populate the band of interest with a fair amount of points. What n is good enough is impossible to say a priori, as one has to do a computer run to get a "feel" for the adequacy of the chosen numbers n, b. For example, to produce a grid with 256 points distributed around 3.5 and follow it with an OPTDesign run, type the following commands.

```
wpts = Grid[Ngrid -> 256,GridSpread -> 3.5];

OPTDesign[p,Center->k0,Radius->r0,GridPoints->wpts];
```

2. *The requirements envelope changes very fast over a frequency region.* This may be the result of your way of setting up the envelope. One possibility is to change it to make it gentler, either by smoothing it or by redefining it. In this case one should verify that these changes preserve the physical requirements for the design. For a successful computer calculation, *one must make sure that the chosen grid has many points in those bands*, as we just described in item 1. The smoothing in OPTDesign runs is specified with the option *Nsmth*. The input *Nsmth→1* indicates a small amount of automatic smoothing; more smoothing is achieved with *Nsmth→5* or *Nsmth→10*. The input *Nsmth→0* corresponds to no automatic smoothing. An example of a command is

```
OPTDesign[p,Center->k0,Radius->r0,Nsmth->5,GridPoints->wpts];
```

3. *Agreement (or lack thereof) between functions as data on a grid and a rational formula.* This topic is important enough to deserve a separate section; we refer the reader to section 5.9.

5.9 Rational Fits

We have seen that the function RationalModel in the package OPTDesign produces rational approximations to data representing an optimal closed-loop transfer function. Rational approximation of functions given as data may be done poorly, or even may go completely wrong for several fundamental reasons (not related to OPTDesign).[4]

---

[4]This is because the problem is one of highly nonlinear optimization. As a consequence,

1. The larger the degree of the denominator, the more likely a rational fit is to converge to a false optimal fit.
2. If data is not reasonably smooth, then fitting routines may give spurious results.
3. Rational fits are very sensitive to initialization.

One may decide whether the rational function chosen to approximate data is of good quality simply by displaying a plot of the absolute value of the difference of the data and the function (evaluated at the frequency values specified by the data) and studying the resulting profile.

The function RationalModel uses an algorithm by L. N. Trefethen for doing Caratheodory-Fejer approximation. To use it, type

```
Trat2 = RationalModel[T,DegreeOfDenominator -> 5]
```

We warn the reader again that RationalModel is not a stand-alone rational fit program and that it must be run in the course of OPTDesign. For a stand-alone Caratheodory-Fejer rational fit on a stable function whose values on a grid *wpts* are given in a variable F, one can use

```
StableFit[F,wpts,DegreeOfDenominator -> 5]
```

The grid should be of the type discussed in section 5.8 and discussed further in section 6.1.1.

The OPTDesign package also contains a more powerful general-purpose routine for approximation of data called *NewtonFit*. See the documentation for NewtonFit in Appendix F.

5.10 Exercises

1. In the problem treated in this chapter, smooth the envelope a lot, say *Smoothing* $\to$ 50, and compare it to the unsmoothed envelope.

2. While fixing all other specs in section 5.2, make the gain-phase margin as small as possible. (This is the Glover–McFarlane [GM90] approach to design.) How much does smoothing affect the answer?

3. Same as problem (2), but vary the tracking error.

For example, the effect of this on RationalModel is that it may not work satisfactorily when

1. *DegreeOfDenominator* $\to n$ is used as input with $n \geq 4$.
2. The numerical error diagnostic in OPTDesign is not small.
3. You least expect it. To be safe you must run with many different initializations.

Part II
More on Design

Chapter 6

Examples

In this chapter we present examples that illustrate the theory developed in previous chapters, as well as the use of some techniques for solving problems in practice.

The design process involves extensive use of computers and algorithms, so some time spent studying their efficient use is beneficial. We discuss briefly in section 6.1 numerical issues that arise in a wide class of algorithms for solving optimization problems. The first control example we solve, the design of a control for a slide drive, is presented in section 6.2.

Many practical situations lead to control design problems with requirements on the time domain functions associated with the system. To use frequency domain methods, one must translate time domain information to the frequency domain. We take a first look at time domain requirements in section 6.3. The requirements for overshoot and settling time and a method for translating information to the frequency domain are introduced in this section. The method is applied in our solution of a design problem in section 6.4.

Finally, in the last section of this chapter we discuss design problems that involve *competing constraints*. By this we mean that two or more constraints are active at the same frequencies, and improving a design with respect to one of them brings a degradation of performance with respect to at least one of the other ones. In section 6.5 we discuss ways to treat competing constraints with our methods, and a design problem is solved using two practical approaches.

6.1 Numerical practicalities

There are several reasons for the numerical difficulties that may arise when solving OPT by computer:

- Points on the $j\omega$ axis for sampling functions are badly chosen.

```
                    . . . . . . . . . . . . . . . . . . . . . . .  .   .

    0.01 b       0.1 b          b        10 b       100 b
```

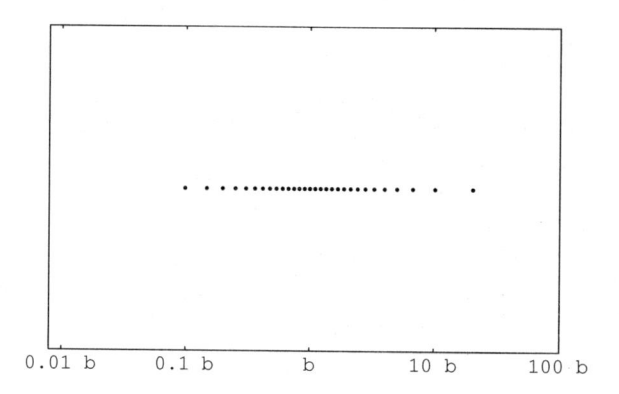

Fig. 6.1. *An example of 64-point grid on the ω axis produced with the OPTDesign package command* `Grid[Ngrid -> n, GridSpread -> b]`, *which gives n points distributed around $\pm b$. Practically all (positive) points lie between $0.1b$ and $10b$.*

- The performance function Γ is not continuous. In the circular performance case, this occurs when the radius function or the center function is not continuous.

- The value of the performance $\Gamma(\omega, T(j\omega))$ is extremely large at some frequencies ω, regardless of the value of T. In circular performance problems, this is the case when the radius gets very close to 0 at one or more frequencies.

- At certain frequency ω, the value of the performance $\Gamma(\omega, T(j\omega))$ does not depend on T.

We discuss these issues below. The emphasis is on circular performance functions, but the comments apply, for the most part, to other performance functions as well.

6.1.1 Sampling functions on the $j\omega$ axis

To solve OPT by computer one must choose a grid of points on the $j\omega$ axis. These points are needed for sampling functions. To have an adequate discrete representation of functions, the grid must contain many points in the frequency bands where the functions being sampled change a lot. For circular performance functions, the center K and the radius R determine which choices of grid points are appropriate. Figures 6.1 and 6.2 illustrate different selections of grids with the software package OPTDesign.[1]

---

[1] The main optimization program, Anopt, works with functions defined on the unit circle instead of the imaginary axis. The linear fractional transformation $s \to b\frac{1+z}{1-z}$ is used to map from the axis to the circle and vice versa. Here b is a constant that the user may set.

6.1. NUMERICAL PRACTICALITIES

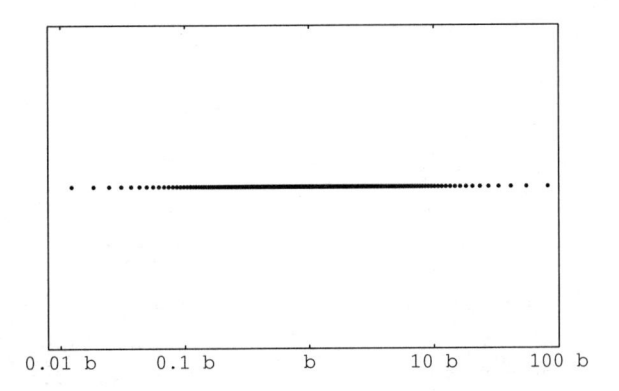

```
0.01 b      0.1 b       b        10 b      100 b
```

Fig. 6.2. *An example of a 512-point grid on the ω axis produced with the OPT-Design package command* Grid[Ngrid $->$ n, GridSpread $->$ b]. *Most of the (positive) points lie between $0.01b$ and $100b$.*

6.1.2 Discontinuous functions

Depending on the way in which requirements are put together, the performance function may not be continuous. For example, the center function may have values that drop from 1 to 0 at a particular frequency. This is undesirable; most algorithms for solving *OPT* have numerical problems when dealing with discontinuous performance functions. Even if the optimal designable function T could be computed, in many cases it wouldn't be acceptable because it is not continuous.

To circumvent this problem in the circular performance case, it is necessary to modify the radius and center functions to make them continuous and even differentiable. Doing this changes the problem being solved, but it still should produce good approximate solutions if handled with care. The trade-off is closeness to the original function versus good numerical properties.

To illustrate a simple way to obtain a continuous function from a discontinuous one, consider the function

$$k(\omega) = \begin{cases} 1 & \text{if } 0 \leq |\omega| \leq 2, \\ 0 & \text{if } 2 \leq |\omega| \leq \infty. \end{cases}$$

The function $k(\omega)$ has a jump discontinuity at $\omega = 2$. We pick a "transition region," say $\omega_a \leq \omega \leq \omega_b$, that contains the location of the jump and interpolate the points $(\omega_a, k(\omega_a))$ and $(\omega_b, k(\omega_b))$ with a linear function $\ell_1(\omega)$. Then we define

$$k_1(\omega) := \begin{cases} k(\omega) & \text{if } 0 \leq |\omega| \leq \omega_a, \\ \ell_1(|\omega|) & \text{if } \omega_a \leq |\omega| \leq \omega_b, \\ k(\omega) & \text{if } \omega_b \leq |\omega| \leq \infty. \end{cases}$$

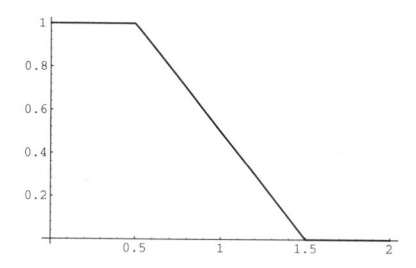

Fig. 6.3. *Plot of the function $k_1(\omega)$ obtained from $k(\omega)$ by using linear interpolation with transition region $1.5 \leq \omega \leq 2.5$. The following Mathematica commands produce $k_1(\omega)$:* `wa = 0.5; wb = 1.5; line1[w_] = InterpolatingPolynomial[wa,1,wb,0,w]; k1[w_] = Which[ Abs[w] <= wa , 1, wa < Abs[w] <= wb , line1[Abs[w]] , wb < Abs[w] , 0];`

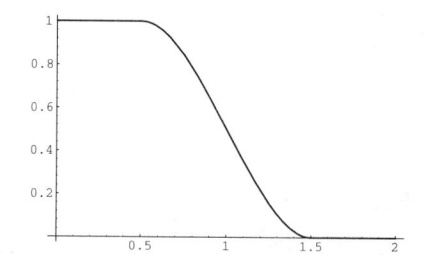

Fig. 6.4. *The plot of the function $k_2(\omega)$ obtained from $k(\omega)$ by using high-order interpolation with transition region $1.5 \leq \omega \leq 2.5$ and derivatives at the endpoints specified to be 0. The following Mathematica commands produce $k_2(\omega)$:* `wa = 0.5; wb = 1.5; poly1[w_] = InterpolatingPolynomial[wa,1,0,wb,0,0,w]; k2[w_] = Which[ Abs[w] <= wa , 1, wa < Abs[w] <= wb , poly1[Abs[w]] , wb < Abs[w] , 0];`

Other types of interpolation are possible. For example, one may do it with a polynomial so that derivatives at the endpoints of the transition region have certain values. See Figs. 6.3 and 6.4.

6.1.3 Vanishing radius function

In some cases, when the radius in circular performance problems vanishes, the mathematics for solving *OPT* break down. However, a vanishing radius is not a difficulty if the problem is set up correctly. Indeed, the radius function *should vanish* at frequency ω_0 whenever $s = j\omega_0$ is a zero of the plant (including infinite frequency). The full justification of this statement requires the theory of Chapter 7, so we explain here only the case where the plant $P(s)$ is stable.

Let $P(s)$ be a strictly proper rational function with no RHP poles. By Theorem 2.3.1 internally stable systems $\mathcal{S}$ with plant $P(s)$ have a closed-loop transfer function of the form

$$T(s) = P(s)T_1(s) \tag{6.1}$$

for some $T_1 \in \mathcal{RH}^\infty$. If the system $\mathcal{S}$ satisfies the closed-loop roll-off inequality

$$T(j\omega)| \leq \alpha |P(j\omega)|,$$

then the latter can be written in terms of T_1 as

$$|T_1(j\omega)| \leq \alpha.$$

That is, we obtain a problem where the variable is T_1 and the radius function is equal to α for all large ω, which is acceptable. Computer programs such as OPTDesign do this step automatically.

6.1.4 Performance function incorrectly defined

The performance function defined by the user should depend on the closed-loop transfer function T at each frequency ω, including $\omega = \infty$.

For example, consider a strictly proper plant $P(s)$ (so that for internally stable systems $\mathcal{S}$, the function T rolls off at infinite frequency), and suppose that we wish to use the magnitude of the sensitivity as our only performance criterion, thus leaving out a requirement on the roll-off. We may write the performance function as

$$\Gamma(\omega, T(j\omega)) = |1 - T(j\omega)|.$$

Since T rolls off, *the performance is equal to 1 at infinite frequency, regardless of the choice of T*. Thus this performance function is not adequate for H^∞ optimization. For details on this see section 10.4 in Part III of [HMer98].

6.2 Design example 1

The system is a slide drive, driven by a DC motor (see Fig. 6.5). This was studied by Brett Larson in his master's thesis [La], written under the guidance of Prof. Fred Bailey. The following description appears in page 3 of the thesis:

> An electrical voltage $u(t)$ drives a DC motor with output speed $\omega_1(t)$. The output of the motor goes through the gear train which steps the speed down to $\omega_2(t)$. The gear train is connected to a table which slides along a rail with the load on top.

The fundamental goal in this problem is to control the speed of the table very precisely.

6.2.1 Electro-mechanical and electrical models

Ignoring the dynamics in the gear train, the problem can be represented by the equivalent circuit in Fig. 6.6.

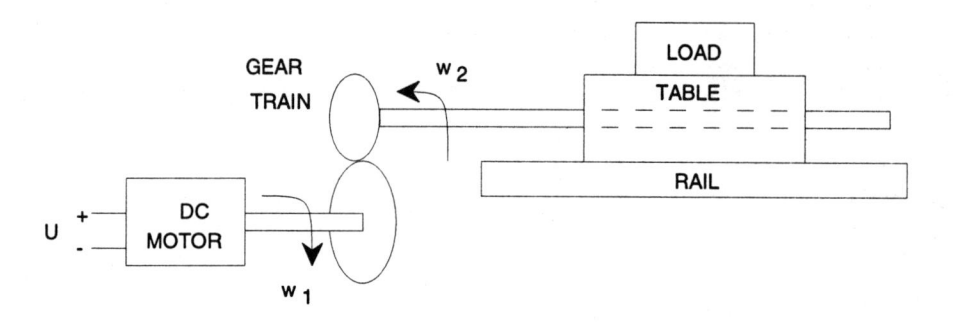

Fig. 6.5. *Slide drive apparatus.*

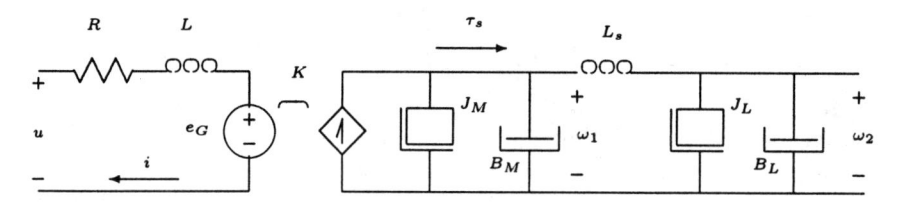

Fig. 6.6. *Electro-mechanical model.*

The following equations correspond to the DC motor.

$$\tau_M = K$$
$$e_G = K\omega_1 \tag{6.2}$$

The parameters for this problem are

$$R = 1\,\Omega,$$
$$L = 5\text{ mH},$$
$$K = 0.2\text{ V/rps} = 0.2\text{ N}-\text{m/amp},$$
$$J_M = 2.0 10^{-3}\text{ kg}-\text{m}^3,$$
$$B_M = 2.0 10^{-3}\text{ N}-\text{m/rps}, \tag{6.3}$$
$$K_s = 1000\text{ N}-\text{m/rad},$$
$$0.5 J_M \leq J_L \leq 5.0 J_M,$$
$$J_{L_{nom}} = J_M,$$
$$B_L = 1\text{ N}-\text{m/rps}.$$

6.2. DESIGN EXAMPLE 1

Figure 6.7 represents this problem with the mechanical parts replaced by their electrical equivalents, where

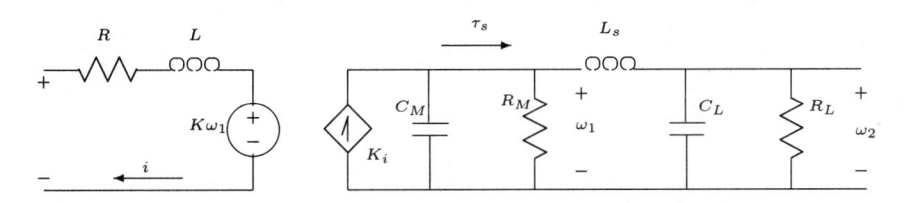

$$
\begin{aligned}
C_m &= J_M \\
R_M &= 1/B_M \\
L_S &= 1/K_s \\
C_L &= J_L \\
R_L &= 1/B_L
\end{aligned}
\tag{6.4}
$$

Fig. 6.7. *Electrical model.*

6.2.2 Mathematical model

Choosing the state vector as

$$
x = \begin{pmatrix} x_1 \\ x_2 \\ x_3 \\ x_4 \end{pmatrix} = \begin{pmatrix} i \\ \omega_1 \\ \tau_x \\ \omega_2 \end{pmatrix},
\tag{6.5}
$$

we have that the system of state equations for the system is

$$
\begin{aligned}
x'(t) &= Ax(t) + bu(t) \\
y(t) &= cx(t) + du(t).
\end{aligned}
\tag{6.6}
$$

Take ω_1 as the output, and scale the frequency by 100 to obtain the following coefficient matrices, where $\alpha = 1/J_L$:

$$A = \begin{pmatrix} -2 & -0.4 & 0 & 0 \\ 1 & -0.01 & -5 & 0 \\ 0 & 10 & 0 & -10 \\ 0 & 0 & \alpha & \alpha \end{pmatrix}$$

$$b = \begin{pmatrix} 2 \\ 0 \\ 0 \\ 0 \end{pmatrix} \tag{6.7}$$

$$c = (0\ 1\ 0\ 0)$$

$$d = (0)$$

The parameter α varies between 1 and 10. The maximum load corresponds to $\alpha = 1$. The plant transfer function from input u to output ω_1 is computed from the formula

$$P_\alpha(s) = c(sI - A)^{-1}b. \tag{6.8}$$

We obtain

$$P_\alpha(s) = \frac{2s^2 + 2\alpha s + 20\alpha}{s^4 + (2.01+\alpha)s^3 + (50.42+12.01\alpha)s^2 + (100+70.52\alpha)s + 104.2\alpha}. \tag{6.9}$$

The parameter α accounts for the variable load in this example. Here α varies between $\alpha = 10$, which is the minimum-load case, and $\alpha = 1$, which is the maximum-load case. The nominal case corresponds to $\alpha = 5$. The nominal plant is

$$\begin{aligned} P_5(s) &= \frac{2s^2 + 10s + 100}{s^4 + 7.01s^3 + 110.47s^2 + 452.6s + 521} \\ &= \frac{(s+2.5 \pm j6.61437)}{(s+2.33827 \pm j0.272238)(s+1.16673 \pm j9.62573)} \end{aligned} \tag{6.10}$$

The nominal plant is stable (so are all the other plants P_α for all $\alpha \in [1, 10]$). The magnitude of the nominal plant is shown in Fig. 6.8.

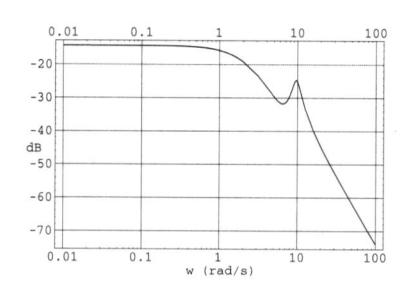

Fig. 6.8. *Magnitude of the nominal plant* $P(s) = P_5(s)$.

6.2. DESIGN EXAMPLE 1

6.2.3 Statement of the problem

The set of requirements given here is taken from [La89].

$$
\begin{aligned}
&\text{a.} && |S(j\omega)| < -20 \text{ dB} && \text{for } |\omega| < 0.12 \text{ rad/s} \\
&\text{b.} && |S(j\omega)| < -10 \text{ dB} && \text{for } |\omega| < 1.0 \text{ rad/s} \\
&\text{c.} && |T(j\omega)| < -20 \text{ dB} && \text{for } |\omega| > 5.0 \text{ rad/s} \\
&\text{d.} && |T(j\omega)| < -40 \text{ dB} && \text{for } |\omega| > 10.0 \text{ rad/s} \\
&\text{e.} && \text{Phase margin is greater than 30 degrees.}
\end{aligned} \qquad (6.11)
$$

According to Larson, these requirements should be met over the full range of load perturbations. We will instead do the (much easier) problem of solving *Design* for the nominal plant only. The problem for loads other than the nominal will not be discussed here.

6.2.4 Reformulation of requirements

Now we rewrite the requirements using the notation of Chapter 3. Observe that the phase-margin constraint needs to be stated as an inequality in terms of the closed-loop function T. Also, note that there is no closed-loop roll-off constraint, so one must be supplied.

Setting a gain-phase margin constraint. To treat the phase margin requirement (6.11e), we rewrite it as a gain-phase margin constraint with $m = 0.5$:

$$|1 - T(j\omega)| < \frac{1}{0.5} = 2 \quad \forall \omega$$

It is clear that the gain-phase margin constraint is the only one that is binding at midrange frequencies $1 < \omega < 5$. Also note that $m = 0.5$ ensures a gain margin of 6 dB (see Fig. 6.9).

Setting a closed-loop roll-off constraint. For many problems with proper compensator one can use the plant $P(s)$ to provide the profile of the envelope, by setting $h(\omega) = |P(j\omega_r)|$ in

$$|T(j\omega)| < \frac{\alpha_r}{h(\omega_r)} h(\omega) \quad \text{for } \omega > w_r \qquad (6.12)$$

In our case we set $\omega_r = 10.0$. A plot of the magnitude of the plant (Fig. 6.8) shows that it has a peak at a frequency close to our choice of ω_r, so near this

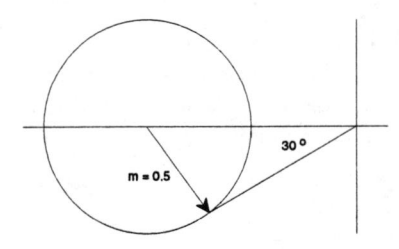

Fig. 6.9. *Derivation of m in design example 1.*

frequency $|P(j\omega)|$ changes quickly. In light of this we do not use the plant in the closed-loop roll-off constraint. Instead, we shall state constraint (6.12) in terms of a function whose magnitude has a simple behavior at frequencies $\omega > \omega_r$, and such that its roll-off corresponds to that of a rational function with relative degree 2. We choose

$$h(\omega) = \left| \frac{1}{(1+j\omega)^2} \right|.$$

The complete set of performance requirements can be stated now:

$$\begin{array}{rclrcl}
|1 - T(j\omega)| & < & 0.12, & 0.0 \leq \omega & \leq 0.1 \\
|1 - T(j\omega)| & < & 0.32, & 0.1 \leq \omega & \leq 1.0 \\
|1 - T(j\omega)| & < & 2, & 1.0 < \omega & < 5.0 \\
|T(j\omega)| & < & 0.1, & 5.0 < \omega & < 10.0 \\
|T(j\omega)| & < & 1.01 \, |1+j\omega|^{-2}, & 10.0 < \omega & < \infty
\end{array} \qquad (6.13)$$

With respect to internal stability we merely note that there are no RHP poles or zeros of the plant $P(s)$. Hence the internal stability of the system with plant $P(s)$ (and proper compensator) is guaranteed by simply requiring that the closed-loop transfer function $T(s)$ have relative degree 2.

A circular performance function

$$\Gamma(\omega, z) = \frac{1}{R(j\omega)^2} |K(j\omega) - z|^2 \qquad (6.14)$$

can be formed from the performance requirements by setting

$$K(j\omega) = \begin{cases} 1, & |\omega| < 5.0 \\ 0, & |\omega| \geq 5.0 \end{cases} \qquad (6.15)$$

and

$$R(j\omega) = \begin{cases} 0.12, & 0.0 \leq |\omega| < 0.1 \\ 0.32, & 0.1 < |\omega| < 1.0 \\ 2.0, & 1.0 < |\omega| < 5.0 \\ 0.1, & 5.0 < |\omega| < 10.0 \\ 1.01 \, |1+j\omega|^{-2} & 0 \leq |\omega| < \infty \end{cases} \qquad (6.16)$$

The center $K(\omega)$ and radius $R(\omega)$ have jump discontinuities (see Figs. 6.10 and 6.11).

The next thing to do is to remove jump discontinuities from the requirements envelope. To do it, we define *continuous* functions $K_1(\omega)$ and $R_1(\omega)$ that approximate $K(\omega)$ and $R(\omega)$ except near jump discontinuities, where $K_1(\omega)$ and $R_1(\omega)$ are defined as linear functions. See Figs. 6.12–6.15.

6.2.5 Optimization

We now have all the elements necessary to state and solve the design problem by optimization of the performance. A run of the program OPTDesign with

6.2. DESIGN EXAMPLE 1

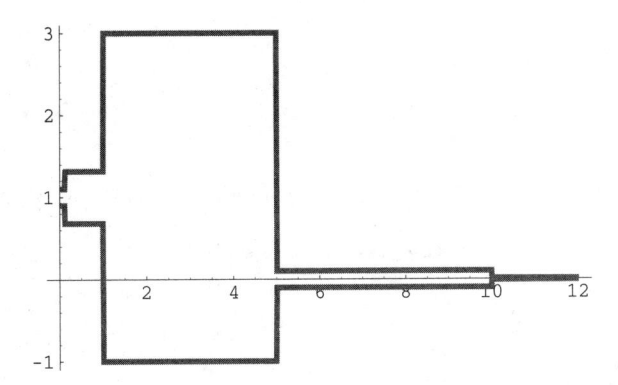

Fig. 6.10. *2-D plot of the requirements envelope determined by the center function $K(\omega)$ and radius function $R(\omega)$.*

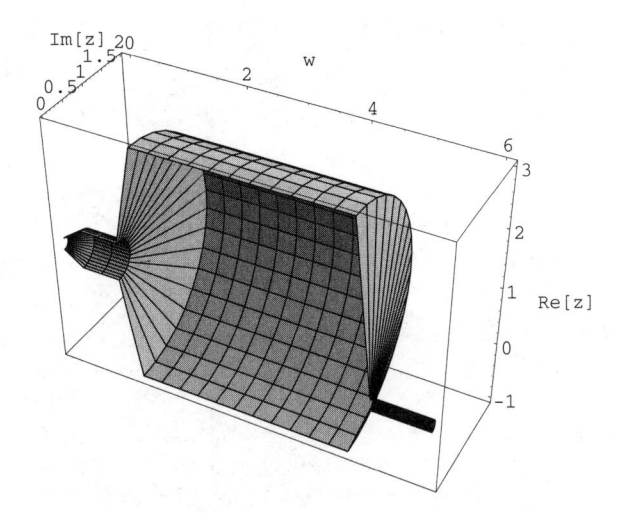

Fig. 6.11. *3-D plot of the requirements envelope determined by the center function $K(\omega)$ and radius function $R(\omega)$.*

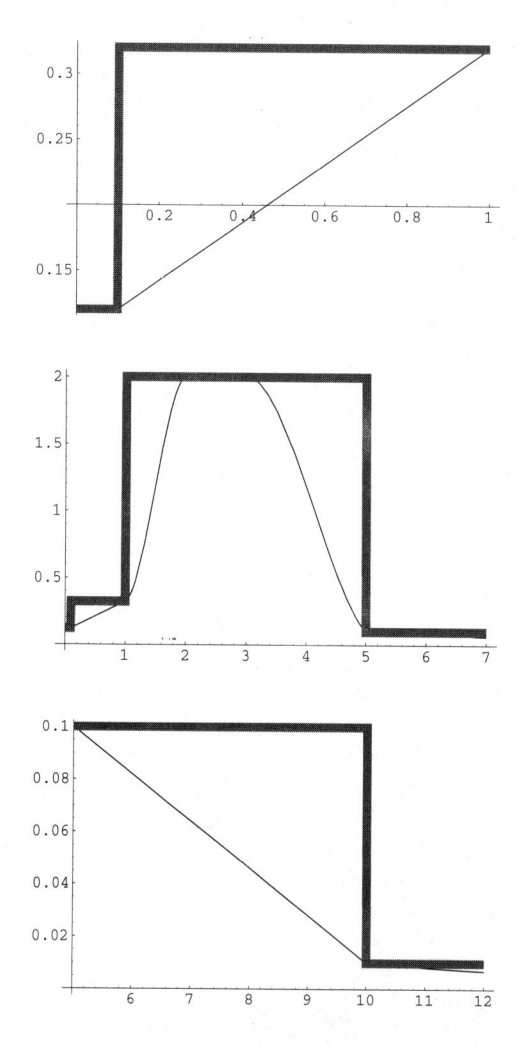

Fig. 6.12. *The radius functions $R_1(\omega)$ (thin) and $R(\omega)$ (thick) on three different frequency ranges.*

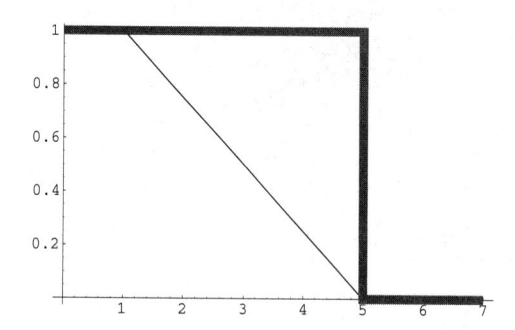

Fig. 6.13. *The center functions $K_1(\omega)$ (thin) and $K(\omega)$ (thick).*

6.2. DESIGN EXAMPLE 1

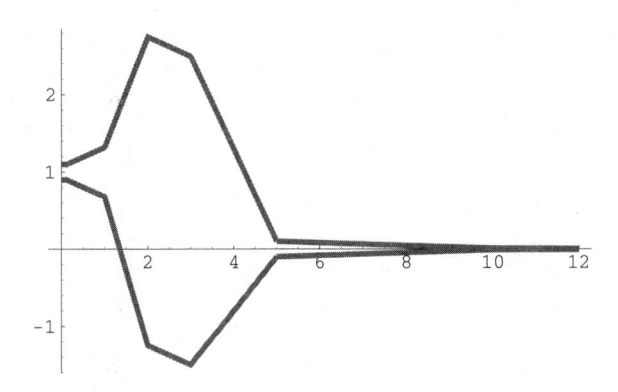

Fig. 6.14. *2-D plot of the requirements envelope corresponding to the center function $K_1(\omega)$ and radius function $R_1(\omega)$.*

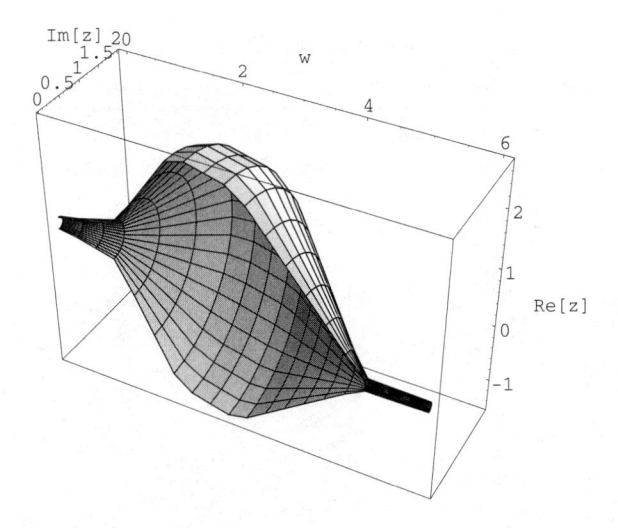

Fig. 6.15. *3-D plot of the requirements envelope corresponding to the center function $K_1(\omega)$ and radius function $R_1(\omega)$.*

256 gridpoints and with heavy smoothing yields $\gamma^* = 0.618$, so there exist solutions to the design problem with the modified envelope (with smoothing). The plots of the optimal closed-loop transfer function T^* and the corresponding sensitivity function $S = 1 - T^*$ indicate that the sensitivity has magnitude -5 dB for frequency ω near 1, while in the original performance requirements it was specified that this magnitude had to be less than -10 dB. See Figs. 6.16–6.18. Thus we must take further action to solve the original design problem.

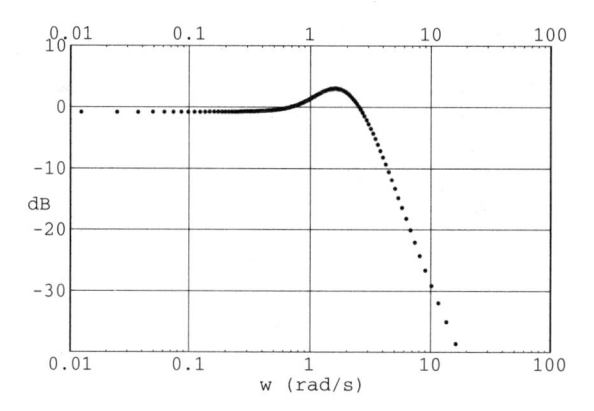

Fig. 6.16. *Bode magnitude plot of T^*.*

6.2.6 Second modification of the envelope and optimization

Note that the function T^* obtained in the previous run has satisfactory characteristics at frequencies other than ω near 1. We now modify the requirements envelope used in the previous run in a way that amounts to tightening the constraints in a small frequency band around the frequency $\omega = 1$. See Figs. 6.19 and 6.20.

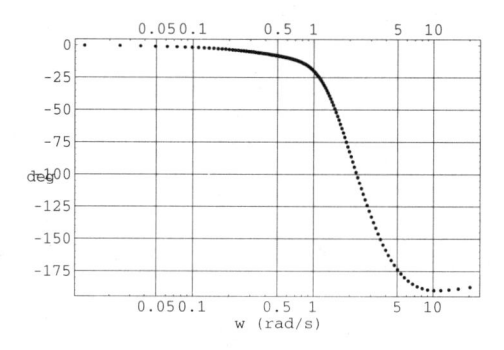

Fig. 6.17. *Bode phase plot of T^*.*

6.2. DESIGN EXAMPLE 1

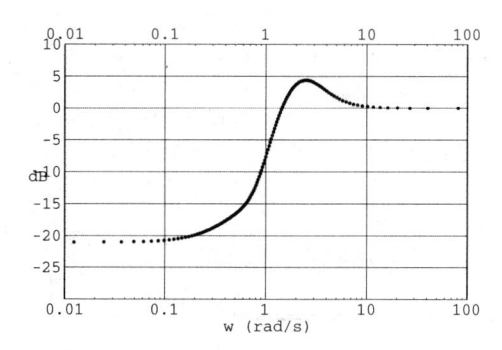

Fig. 6.18. *Bode magnitude plot of the sensitivity function $S^* = 1 - T^*$.*

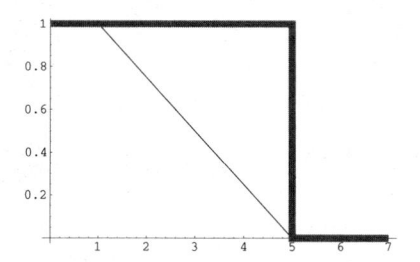

Fig. 6.19. *The center function $K_2(\omega)$ values drop in magnitude to the right of $\omega = 1.3$. Here $K_2(\omega)$ is shown as the thin line and $K(\omega)$ is shown as the thick line.*

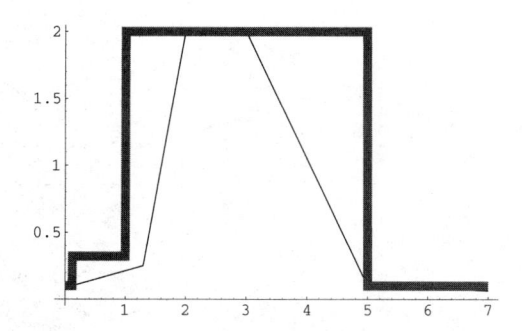

Fig. 6.20. *The radius function $R_2(\omega)$ values are smaller than 0.25 for $0 \leq \omega \leq 1.3$. Here $R_2(\omega)$ is shown as the thin line and $R(\omega)$ is shown as the thick line.*

The run of OPTDesign with the new envelope produces an acceptable closed-loop transfer function. A degree 6 rational approximation to the new optimal closed-loop transfer function is readily obtained with the function RationalModel. We have

$$T_2(s) = (2.0+s)^{-2} + \frac{14.4095+28.535\,s+24.2206\,s^2+7.29848\,s^3+0.449194\,s^4}{(2.0+s)^2\,(5.47076+5.42235\,s+3.77192\,s^2+1.13201\,s^3+0.202961\,s^4)} \quad (6.17)$$

The compensator corresponding to $T_2(s)$ is

$$C_2(s) = \frac{0.5\left(19.8803+33.9574\,s+27.9925\,s^2+8.43049\,s^3+0.652155\,s^4\right)}{(50.0+5.0s+1.0s^2)}$$

$$\times \frac{\left(521.0+452.6\,s+110.47\,s^2+7.01\,s^3+s^4\right)}{(2.00274+9.61507\,s+14.2554\,s^2+16.6076\,s^3+8.45965\,s^4+1.94385\,s^5+0.202961\,s^6)} \quad (6.18)$$

See Figs. 6.21–6.25.

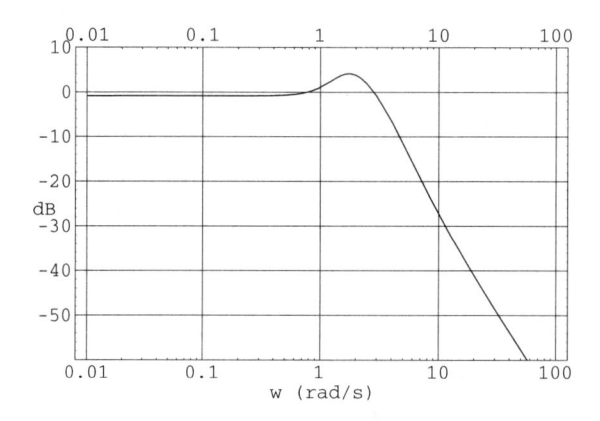

Fig. 6.21. Bode magnitude plot of $T_2(s)$.

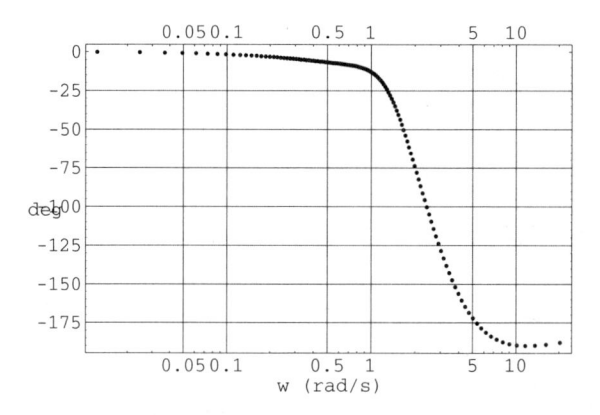

Fig. 6.22. Bode phase plot of $T_2(s)$.

6.3. TIME DOMAIN PERFORMANCE REQUIREMENTS

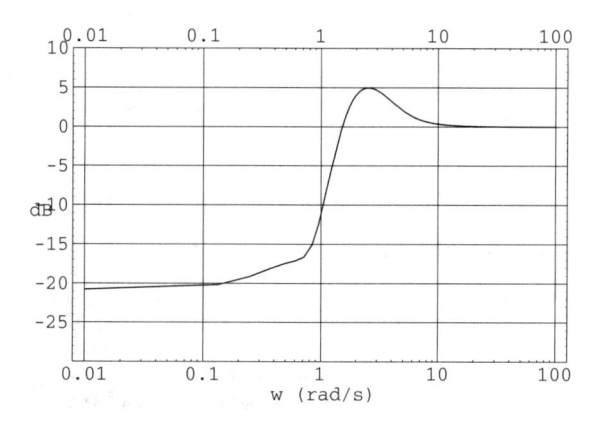

Fig. 6.23. *Bode magnitude plot of the sensitivity function $S_2(s) = 1 - T_2(s)$.*

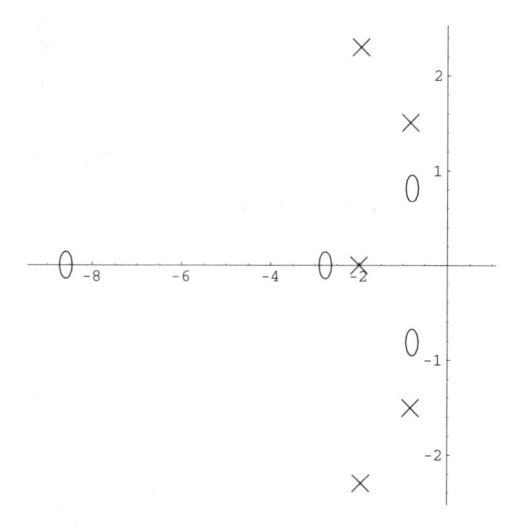

Fig. 6.24. *Plot of zeros and poles of the closed-loop function $T_2(s)$.*

Conclusion. A numerical solution to the original design problem with nominal plant has been found. Plots of the magnitude and phase of the (discrete) optimal compensator are shown in Figs. 6.26 and 6.27.

6.3 Time domain performance requirements

Time domain requirements on the closed-loop system $\mathcal{S}$ must be translated to the frequency domain in order to solve design problems with the methods used in this book. In this section, two time domain requirements are introduced, and a technique for treating them is given. The technique is applied later in a second design problem (section 6.4).

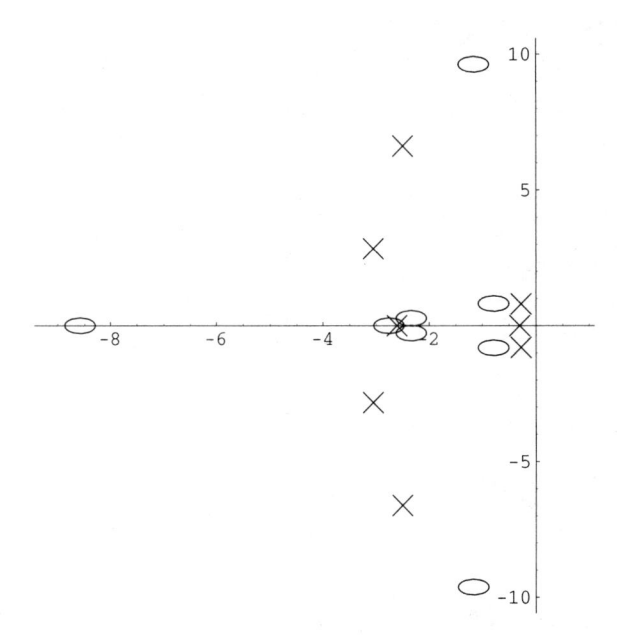

Fig. 6.25. *Plot of zeros and poles of the compensator* $C_2(s) = \frac{1}{p(s)} \frac{T_2(s)}{1-T_2(s)}$.

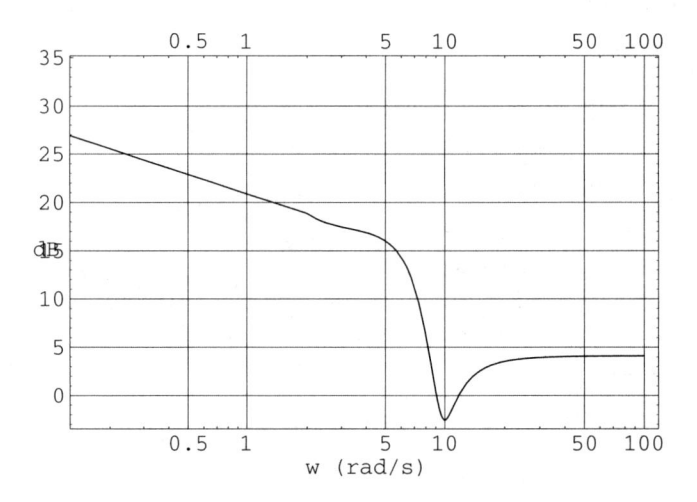

Fig. 6.26. *Magnitude of the optimal compensator.*

6.3. TIME DOMAIN PERFORMANCE REQUIREMENTS

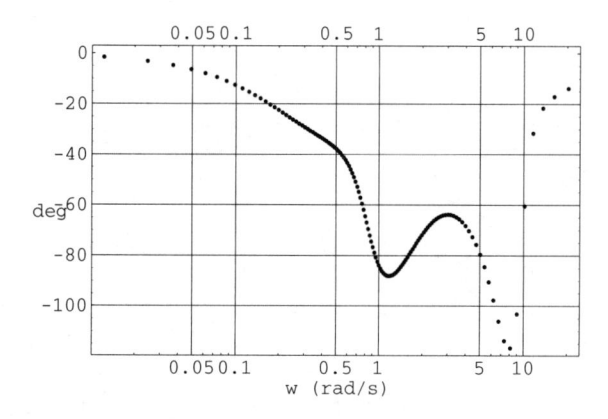

Fig. 6.27. *Phase of the optimal compensator.*

6.3.1 Two common time domain requirements

Two common time domain requirements used for system design are the overshoot requirement and the settling time requirement. Both are defined in terms of the step response curve $c(t)$, which is the response of the system in the time domain to an input that is a unit step (see Fig. 6.28). The step response function has $T(s)/s$ as Laplace transform, i.e.,

$$c(t) = \mathcal{L}^{-1}[T(s)/s](t). \tag{6.19}$$

The requirements we present in this sections are as follows.

- *Overshoot requirement*: Given $M > 0$,

$$c(t) - 1 < M \quad \text{for all } t > 0. \tag{6.20}$$

- *Settling time requirement*: Given $t_s > 0$ and $\Delta > 0$,

$$|c(t) - 1| < \Delta \quad \text{for all } t > t_s. \tag{6.21}$$

6.3.2 A naive method

A way to handle time domain requirements in the frequency domain is to use a low-order rational function T_{ref} as a reference. The idea is to choose, before the design process begins, a function T_{ref} that is the transfer function of a well-understood system that meets the time domain requirements. Then frequency domain requirements are established by using plots of the function T_{ref}. The hope is that systems that satisfy these frequency domain requirements will also satisfy the original time domain requirements.

A choice of closed-loop reference function that is usually mentioned in control books is

$$T_{\text{ref}}(s) = \frac{\omega_n^2}{s^2 + 2\zeta s \omega_n + \omega_n^2}. \tag{6.22}$$

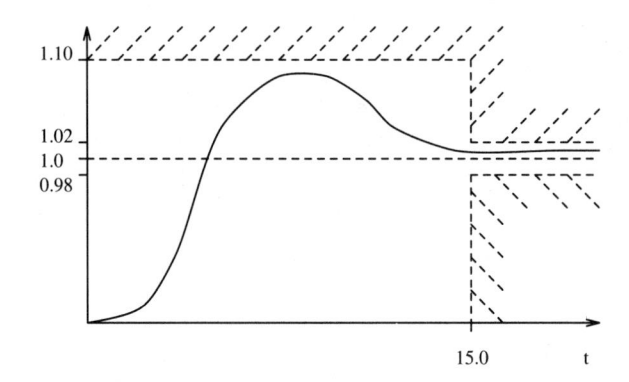

Fig. 6.28. *A step response function that satisfies the overshoot and settling time requirements when $M = 0.1$, $t_s = 15.0$, and $\Delta = 0.02$. In this figure, the step response curve cannot run into the shaded area without violating the constraints.*

This system is known to satisfy the relations (see [FPE86])

$$t_s \approx \frac{4.6}{\zeta \omega_n} \qquad (6.23)$$

and

$$M_p = \exp\left(\frac{-\zeta \pi}{\sqrt{1-\zeta^2}}\right). \qquad (6.24)$$

Hence $T_{\text{ref}}(s)$ is completely determined by specifying t_s and M_p. Plots of functions T_{ref} and $1 - T_{\text{ref}}$ are then used to determine frequency domain parameters (bandwidth, tracking, etc).

The advantage of (6.22) is that because of its low order, it is very easy to generate parameters ω_n and ζ to suit one's needs. However, the strength of (6.22) is also its weakness. This is a consequence of the initial value and final value theorems [C44] [LP61], which (very roughly) have the following rule of thumb as a corollary.

> TIME DOMAIN VS. FREQUENCY DOMAIN RULE. *For internally stable systems with type n plant ($n \geq 0$), high-frequency behavior of the closed-loop transfer function determines the behavior of the step response function near time zero, and very low frequency behavior of the closed-loop transfer function determines the behavior of the step response function for large times. The same is true for ramp response when $n \geq 1$ and parabola response when $n \geq 2$.*

It is a fact that functions $T_{\text{ref}}(s)$ given by (6.22) come from systems with type $n = 1$ plant (i.e., it has a simple pole at $s = 0$). Hence if settling time and even overshoot are important in a design problem, then the function T_{ref} given in (6.22) may give useful information *only* if the plant of the system to be designed is of type $n = 1$.

6.3.3 A refinement of the naive method

We now describe a way of treating time domain requirements in design problems where the plant of the system is of type n.

Suppose that we have a type n plant and that we wish to solve a design problem so that in particular requirements on the step response and overshoot are satisfied.

The internal stability requirement forces the following interpolation constraints on the closed-loop function $T(s)$ (see section 3.4.6):

$$T(0) = 1, \quad T'(0) = 0, \quad ,\ldots, \quad ,T^{(n-1)}(0) = 0 \qquad (6.25)$$

It is easy to see that if $T(s)$ is given by the formula below,[2] then $T(s)$ has relative degree 1 and satisfies (6.25).

$$T(s) = 1 - \frac{s^n}{(1+s)^n} + \frac{s^n}{(1+s)^{n+1}} T_1(s), \qquad (6.26)$$

where $T_1(s)$ is any bounded and stable function. We will pick $T_{\text{ref}}(s)$ so that it satisfies equation (6.26), but to make the task of choosing T_{ref} manageable, we will consider only those T given by

$$T(s) = 1 - \frac{s^n}{(1+s)^n} + \frac{s^n}{(1+s)^{n+1}} \frac{c(s+a)}{s+b}, \qquad (6.27)$$

where $b > 0$. The real parameters a, b, and c are chosen by trial and error so that

$$s_{\text{ref}}(t) = \mathcal{L}^{-1}\left(\frac{T_{\text{ref}}(s)}{s}\right) \qquad (6.28)$$

satisfies overshoot and settling time requirements.

The formula for a family of reference functions given by (6.27) may be altered to include other time domain considerations. For example, if one would like to require that the step response have very little change for an initial time interval, then one is forced to consider closed-loop functions with a relative degree that is greater than or equal to that of the plant. Also, the magnitude of $T_{\text{ref}}(s)$ for large frequencies is important. To obtain formulas for such families of functions $T_{\text{ref}}(s)$ the reader may use the theory of interpolation developed in Chapter 7.

6.4 Design example 2

The following design problem is taken from [FPE86], p. 497. *Design* is treated here with the naive method for converting time domain requirements described in section 6.3.2.

---

[2] The reader interested in a derivation of this and similar formulas should refer to Chapter 7.

6.4.1 Statement of the problem

The problem is to design a satellite's attitude control. The plant is the rational function
$$P(s) = \frac{0.036(s+25)}{s^2(s+0.02+j)(s+0.02-j)}.$$
The requirements on the step response function $c(t)$ are overshoot
$$c(t) < 1.15 \quad \text{for all} \ \ t \geq 0 \tag{6.29}$$
and settling time
$$|c(t) - 1| < 0.01 \quad \text{for all} \ \ t \geq 14.0. \tag{6.30}$$
We need to find a closed-loop system $\mathcal{S}$ that is internally stable and meets the performance requirements.

6.4.2 Translation of time domain requirements

The plant function $p(s)$ has an order 2 pole at $s = 0$; that is, the plant is of type $n = 2$. Since the behavior of the system at initial times is not an issue, to choose a reference closed-loop function we may use formula (6.27).

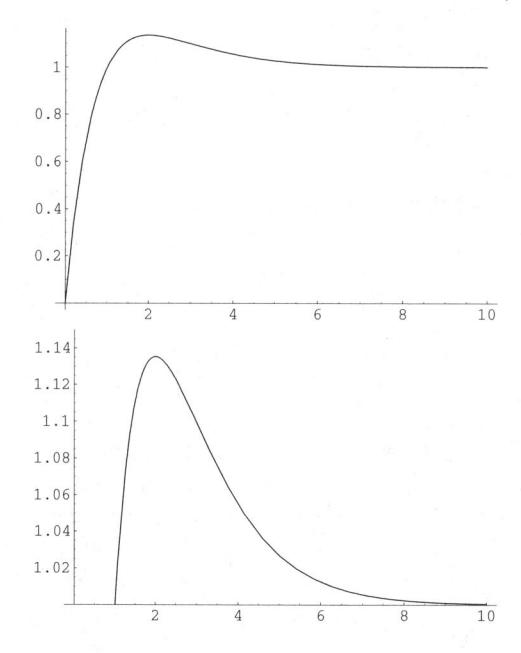

Fig. 6.29. *Step response for a system with closed-loop function T_{ref}.*

Although there are many choices of a, b, and c, the value $c = 0$ gives reasonable step response (Fig. 6.29). Hence we set
$$T_{\text{ref}}(s) = \frac{1+2s}{(1+s)^2} \tag{6.31}$$

6.4. DESIGN EXAMPLE 2

Due to the pole of the plant at $s = 0$, the plant bound constraint is binding at low frequencies. A plot of the closed-loop plant for the function $T_{\text{ref}}(s)$ is shown in Fig. 6.30, while the magnitude of $T_{\text{ref}}(s)$ and of the sensitivity function $S(s) = 1 - T_{\text{ref}}(s)$ are depicted in Fig. 6.31. Figures 6.30 and 6.31 suggest a

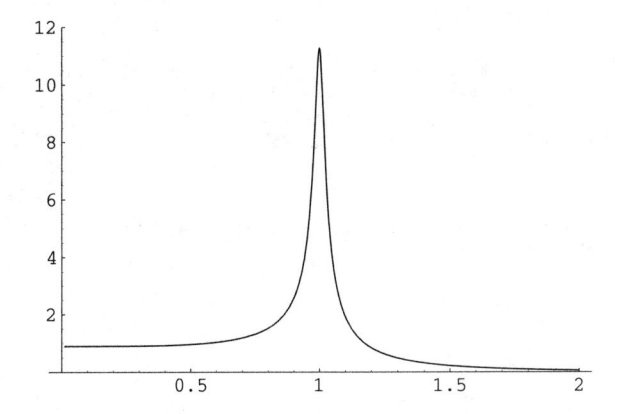

Fig. 6.30. *Plot of the plant bound function for the system when the closed-loop function it T_{ref}.*

center function $k(j\omega)$ that is equal to 1 at low frequencies (up to $w_p = 0.7$, say) and then drops to be close to 0 after $w_b = 2.0$. We choose k to be piecewise linear, so that if $\ell(\omega)$ denotes a line interpolating $(0.7, 1)$ and $(2.0, 0)$, then

$$k(\omega) := \begin{cases} 1 & 0 \leq \omega \leq 0.7, \\ \ell(\omega) & 0.7 < \omega \leq 2.0, \\ 0 & 2.0 < \omega \leq \infty. \end{cases} \qquad (6.32)$$

The radius function has the following characteristics. At low frequencies

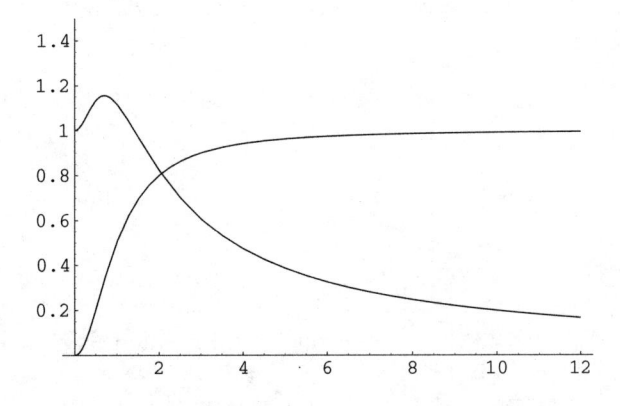

Fig. 6.31. *Magnitudes of $T = T_{\text{ref}}$ and of $S = 1 - T_{\text{ref}}$.*

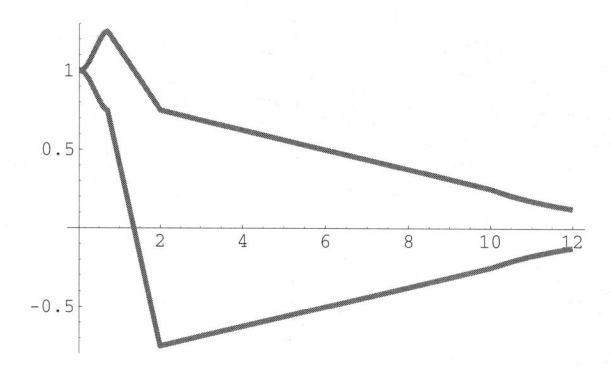

Fig. 6.32. *2-D plot of the requirements envelope.*

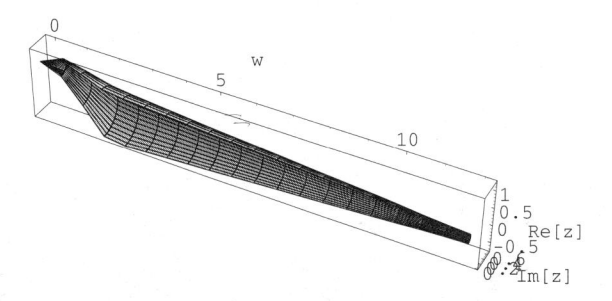

Fig. 6.33. *3-D plot of the requirements envelope.*

($\omega < \omega_p$), the radius equals $\alpha_p/p(j\omega)$, where $\alpha_p = 0.9$ is taken from Fig. 6.30. At frequency $\omega_b = 2$, Fig. 6.31 suggests a radius equal to 0.75. Finally, we pick the roll-off frequency to be $\omega_r = 10.0$. Figure 6.31 suggests a radius function with value 0.25 at the roll-off frequency $\omega = \omega_r = 10.0$, so we set the radius to linearly interpolate these values. For higher frequencies, the radius is set to a multiple of the magnitude of the plant, $|p(j\omega)|$. Let $\ell_1(\omega)$ (resp., $\ell_2(\omega)$) denote the linear interpolant between frequencies $\omega = 0.7$ and $\omega = 2.0$ (resp., $\omega = 2.0$ and $\omega = 10.0$). Then $r(\omega)$ is given by

$$r(\omega) := \begin{cases} \frac{0.90}{|p(j\omega)|} & 0 \le \omega \le 0.7, \\ \ell_1(\omega) & 0.7 < \omega \le 2.0, \\ \ell_2(\omega) & 2.0 < \omega \le 10.0, \\ 0 & 2.0 < \omega \le \infty. \end{cases} \qquad (6.33)$$

See Figs. 6.32 and 6.33. Computer code for this example is given in Appendix C and in the file appendixch6.nb.

A computer run with OPTDesign produces an optimal value $\gamma^* = 1.46$, so there is no solution to the problem with the modified envelope requirements.

6.4. DESIGN EXAMPLE 2

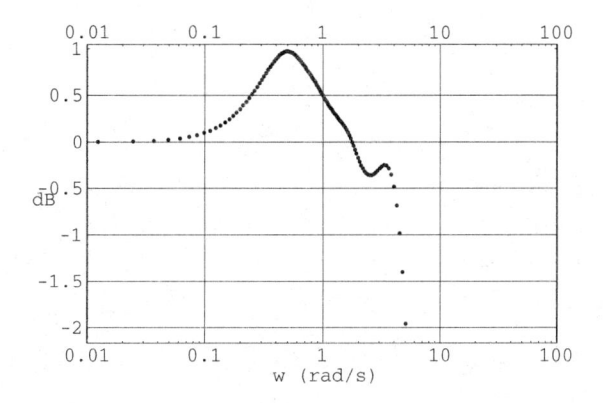

Fig. 6.34. *Bode magnitude plot of T^*.*

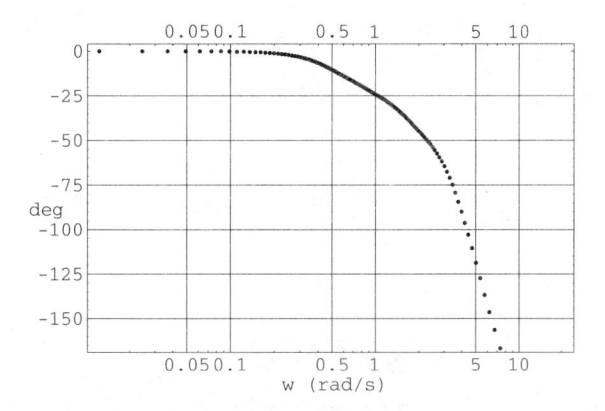

Fig. 6.35. *Bode phase plot of T^*.*

OPTDesign produces an optimal function T^* anyway, that is, a function that comes closest to meeting all the constraints. See Figs. 6.34 and 6.35 for the Bode plots of T^*. The function T^* may be useful for design purposes, since the correspondence between frequency domain and time domain requirements is not explicitly known and something was lost in the (very rough) translation we performed.

A rational approximation $T_1(s)$ of the (discrete) function T^* is readily obtained with the tools of the package OPTDesign. We get

$$T_1(s) = \frac{16.+32.0s}{(2.0+s)^4} + \frac{s^2\left(45.3565+78.823s+11.0016s^2\right)}{(2.0+s)^5\left(3.31431+0.614396s+0.0712923s^2\right)} \tag{6.34}$$

The step response function $\text{step}(t) = \mathcal{L}^{-1}(T(s)/s)$ is shown in Figs. 6.36 and 6.37. The compensator C_1 that corresponds to $T_1(s)$ has high order. Its zeros

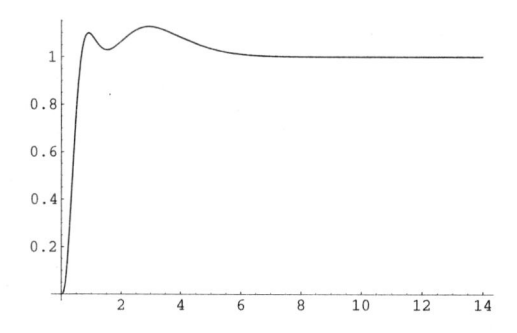

Fig. 6.36. *Step response function for closed-loop T_1.*

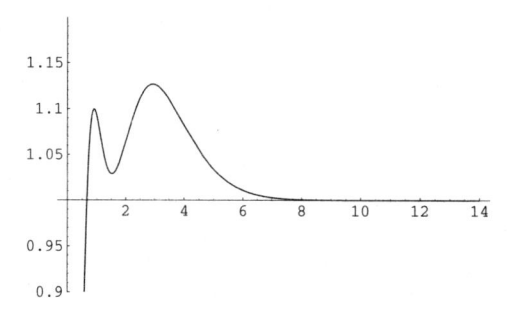

Fig. 6.37. *Another view of the step response function for closed-loop T_1.*

and poles are shown in Figs. 6.38 and 6.39.

$$C_1(s) = \frac{2947.23 + 8032.27s + 8899.41s^2 + 11031.9s^3 + 6119.53s^4 + 2908.84s^5 + 368.97s^6}{2843.26 + 2194.74s + 1336.73s^2 + 357.889s^3 + 45.493s^4 + 3.10963s^5 + 0.0712923s^6}. \tag{6.35}$$

The zeros and poles of $C_1(s)$ are

$$\mathcal{Z}(C_1) = \{-5.8096, -0.759911 \pm 1.44744j, -0.514252, -0.02 \pm 1.0j\} \tag{6.36}$$

$$\mathcal{P}(C_1) = \{-25.0, -5.93595 \pm 6.76924j, -5.26811, -0.738992 \pm 1.78597j\} \tag{6.37}$$

See Figs. 6.38 and 6.39.

Order reduction techniques may be applied to obtain a compensator with lower order and satisfactory characteristics. Here we shall try something very simple that yields a lower-order compensator that may be acceptable. By inspection, we locate pole-zero pairs of $C_1(s)$ that are close and proceed to cancel them. Of course, this produces changes in the frequency response of $C_1(s)$, but the hope is that these changes are not too large (we are partially justified in expecting this because canceled pairs are "close" to a perfect cancellation). By proceeding with cancellation as explained above, we obtain the compensator $C_2(s)$ given by

$$C_2(s) = \frac{1.03657(1+(0.019992-0.9996i)s)(1+(0.019992+0.9996i)s)(1+1.94457s)}{(1+0.04s)(1+(0.0732308-0.083511i)s)(1+(0.0732308+0.083511i)s)} \tag{6.38}$$

6.4. DESIGN EXAMPLE 2

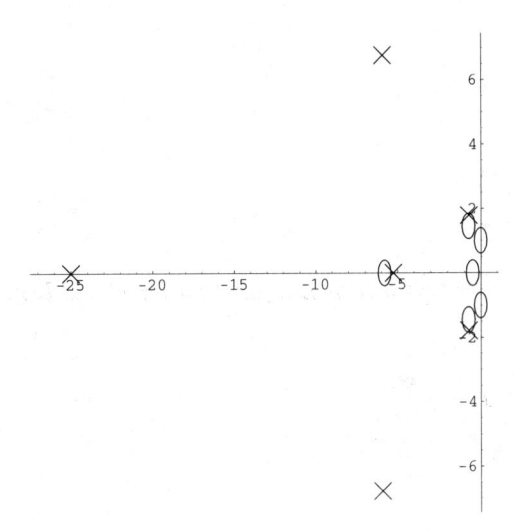

Fig. 6.38. *Plot of zeros and poles of $C_1(s)$.*

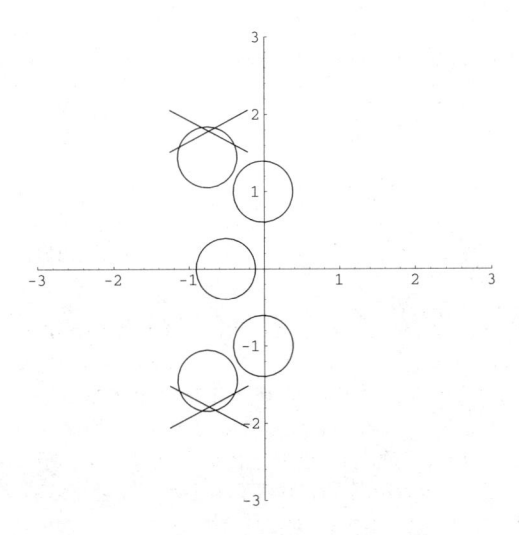

Fig. 6.39. *A view of the zeros and poles of $C_1(s)$ close to the origin.*

The zeros and poles of $C_2(s)$ are shown in Fig. 6.40, while a plot of the magnitudes of $C_1(s)$ and $C_2(s)$ is given in Fig. 6.41.

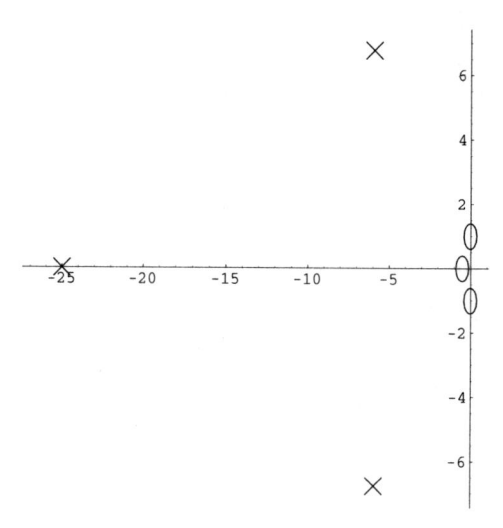

Fig. 6.40. *Zeros and poles of $C_2(s)$.*

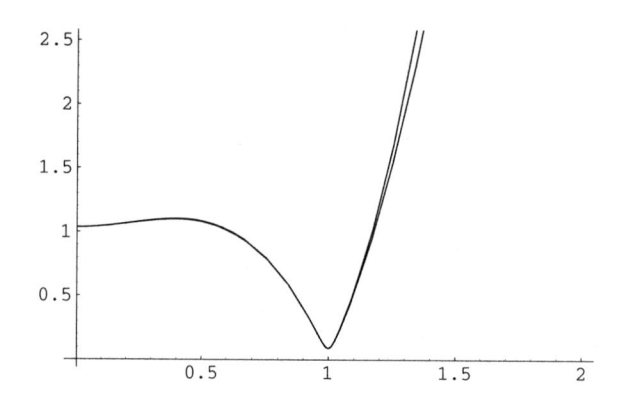

Fig. 6.41. *Magnitudes of $C_1(s)$ and $C_2(s)$.*

The closed-loop function $T_2(s)$ that comes from the compensator $C_2(s)$ is given by

$$T_2(s) = \frac{0.0313827(1+(0.019992\pm0.9996j)s)\cdots}{0.784568+1.77664s+2.1172s^2+2.11976s^3+1.31856s^4+\cdots}$$

$$\frac{\cdots(1+(0.0590236\pm0.0106008j)s)(25+s)(1+2.06645s)}{\cdots+0.3299s^5+0.0432062s^6+0.00363698s^7+0.000160706s^8+2.656410^{-6}s^9}$$
(6.39)

Finally, the step response that corresponds to the choice of $C_2(s)$ as compensator is shown in Fig. 6.42.

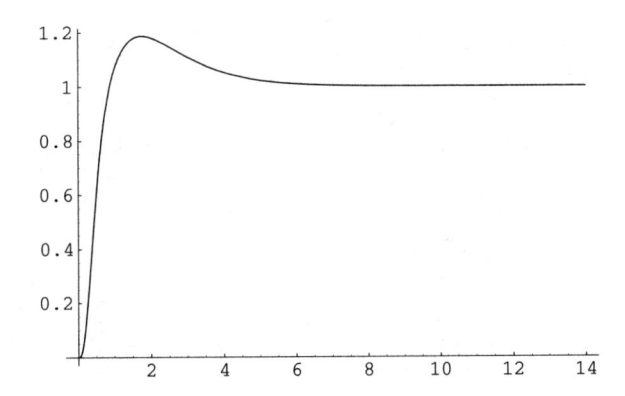

Fig. 6.42. *Step response corresponding to the closed-loop function $T_2(s)$.*

Conclusion. We obtained an order 6 rational compensator so that the corresponding closed-loop system has settling time of 9s and overshoot of 15%. We also obtained an order 3 compensator with overshoot of 20% and settling time of 10s. It is conceivable that the overshoot may be lowered, as there is some room for improvement (by letting the settling time increase).

6.5 Performance for competing constraints

Until now we have dealt with design problems of the circular type; that is, the performance envelope has the shape of a circle at each frequency point. However, in practical design it is common to encounter problems with more than one valid performance requirement on a single frequency band. We say that such design problems have *competing constraints*.

To discuss the issues involved in solving design problems with competing constraints we will focus on the reference problem ($Prob_1$), stated below. While ($Prob_1$) is not completely general, it is all we need to understand in order to be able to attack more difficult problems.

($Prob_1$)   Given a plant $P(s)$ and nonnegative functions $W_1(\omega)$ and $W_2(\omega)$, find $T \in \mathcal{RH}^\infty$ so that $\mathcal{S}$ is internally stable and

$$W_1(\omega)\,|\,1 - T(j\omega)\,| \leq 1 \quad \text{for all } \omega \qquad (6.40)$$
$$W_2(\omega)\,|\,T(j\omega)\,| \leq 1 \quad \text{for all } \omega. \qquad (6.41)$$

For simplicity and physical realism we shall assume that $W_1(\omega)$ is near zero or is zero at very high frequencies, that $W_2(\omega)$ is near zero or is zero at very low frequencies, and that $W_1(\omega)$ has magnitude comparable to $|P(j\omega)|^{-1}$ for large frequencies. In particular, we do not impose a relative degree requirement on the compensator.

The next two subsections describe practical approaches to solving ($Prob_1$).

6.5.1 Rounding corners of performance functions

If the frequency ω is a fixed real number, then the set $\mathcal{S}_\omega$ of all possible complex numbers T that satisfy both inequalities in (Prob$_1$) is a subset of the complex plane. The set $\mathcal{S}_\omega$ (shown in Fig. 6.43) can be described in symbols as

$$\mathcal{S}_\omega = \{z:\ W_1(\omega)\,|1-z|\ \leq\ 1 \text{ and } W_2(\omega)\,|z|\ \leq\ 1\,\}$$

or

$$\mathcal{S}_\omega = \{z:\ \max(\,W_1(\omega)\,|1-z|,\ W_2(\omega)\,|z|\,)\ \leq\ 1\,\}$$

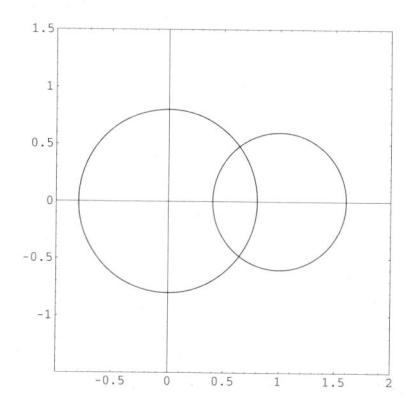

Fig. 6.43. *Sublevel sets $\mathcal{S}_\omega$ arising from two circular performance requirements: $|1-T| < 0.6$ and $|T| < 0.8$. The intersection of the two sets has corners in its boundary.*

Note that $\mathcal{S}_\omega$ is the intersection of two disks, so it typically has "corners" in its boundary. As a consequence of this, the requirements envelope cannot be expressed in terms of center and radius functions like the examples we have encountered so far in this book.

Performance functions with corners in the boundary of the level sets are not differentiable and are difficult to treat numerically. In this section we shall describe two approaches to dealing with (Prob$_1$) numerically, but first we restate (Prob$_1$) as

(Prob$_2$)   Given a plant $P(s)$, nonnegative functions $W_1(\omega)$ and $W_2(\omega)$, and

$$\Gamma(\omega, z) = \max\{W_1(\omega)\,|1-z|,\ W_2(\omega)\,|z|\,\} \tag{6.42}$$

find T that minimizes $\gamma(T) = \sup_\omega \Gamma(\omega, T(j\omega))$ over all T that make $\mathcal{S}$ internally stable.

Mathematically speaking, (Prob$_2$) is what one must solve in order to solve (Prob$_1$). However, (Prob$_2$) is difficult numerically. One practical approach to

6.5. PERFORMANCE FOR COMPETING CONSTRAINTS

solving (Prob$_2$) is to compromise by replacing the performance function Γ in (6.42) by a (numerically) better-behaved function,

$$\Gamma_p(\omega, z) = W_1(\omega)^p \, | \, 1 - z \, |^p \; + \; W_2(\omega)^p \, | \, z \, |^p, \qquad (6.43)$$

where p is an integer such as 2, 4, or 8. Pictorially, doing this corresponds to "rounding the corners of the sublevel sets" to obtain a more gentle performance function. For example, $p = 2$ rounds the corners of the sublevel sets of the performance function (6.42) tremendously,[3] while $p = 8$ does less rounding but is more demanding numerically. See Fig. 6.44.

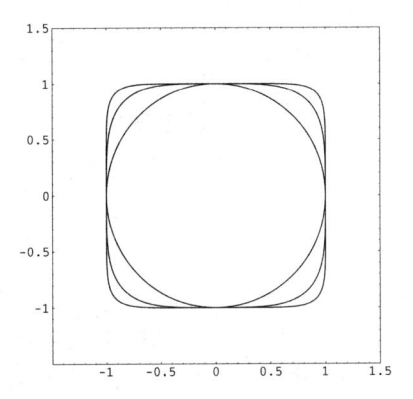

Fig. 6.44. *Plot of level sets* $\mathcal{S}_\omega = \{x + jy : x^p + y^p \leq 1\}$, *for* $p = 2$ *(disk)*, $p = 4$, *and* $p = 8$.

Example. Consider $P(s) = \frac{1}{s+1}$, $W_1 = |P(j\omega)|$, and $W_2 = |P(j\omega)|^{-1}$ in (Prob$_1$). Optimization with the performance function Γ_p, for $p = 2, 4, 8$, initial guess $T^0(s) = 1$, and 32 gridpoints yields (after model reduction) an optimal function $T^p(s)$ where

$$T^2(s) \;=\; \frac{0.716679 \,(1+s)\,\left(1.72022 + 2.38562s + 1.0s^2\right)}{2.95307 + 5.81557s + 4.1023s^2 + 1.0s^3},$$

$$T^4(s) \;=\; \frac{0.613209 \,(1+s)\,\left(9.85094 + 7.1585\,s + 1.0s^2\right)}{14.1019 + 19.2524s + 8.44469s^2 + 1.0s^3}$$

$$T^8(s) \;=\; \frac{0.547494 + 0.991601s + 0.510833s^2}{1.24365 + 1.85885s + 0.897497s^2}$$

Table 6.1 gives information about other output, and Figs. 6.45–6.47 display plots of relevant functions.

Conclusion. Solutions T^p for (Prob$_1$) were obtained using performance functions Γ_p, for $p = 2, 4, 8$. The function $T^8(s)$ gives the best overall weighted sensitivity and weighted magnitude (in more difficult examples its calculation

---

[3] A square root in the formula of the function Γ_2 has been removed, since for optimization purposes such power affects the *optimal value* of the performance, but the function that optimizes the performance does not change.

Table 6.1. *Computer output: Rounding corners method.*

Perf. fn.	$\sup_\omega \Gamma(\cdot, T)$	Flat	GrAlign	$\sup_\omega |W_1(1-T)|$	$\sup_\omega |W_2 T|$
$\Gamma_2(\omega, z)$	0.51	0.0002	0	0.58	0.72
$\Gamma_4(\omega, z)$	0.14	0.0011	0	0.57	0.61
$\Gamma_8(\omega, z)$	0.01	0.0036	0	0.56	0.57

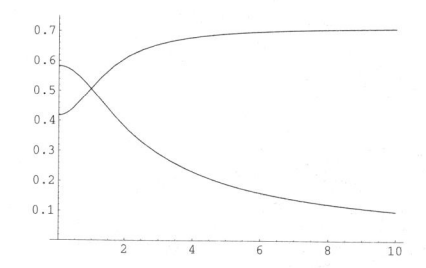

Fig. 6.45. *Plot of* $W_1(\omega)\left|1 - T^2(j\omega)\right|$ *and* $W_2(\omega)\left|T^2(j\omega)\right|$.

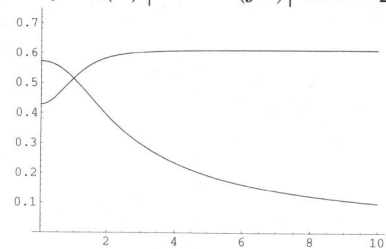

Fig. 6.46. *Plot of* $W_1(\omega)\left|1 - T^4(j\omega)\right|$ *and* $W_2(\omega)\left|T^4(j\omega)\right|$.

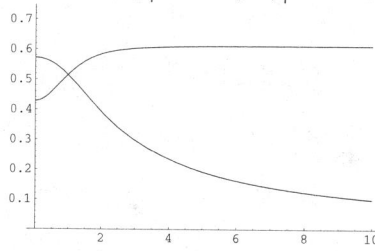

Fig. 6.47. *Plot of* $W_1(\omega)\left|1 - T^8(j\omega)\right|$ *and* $W_2(\omega)\left|T^8(j\omega)\right|$.

could be complicated by the appearance of numerical noise in the computations). Of the three computed answers, $T^2(s)$ is the one that has the least satisfactory overall weighted sensitivity and weighted magnitude, but its calculation entailed little numerical difficulties. Finally, $T^4(s)$ is an intermediate case. Computer code for this example is given in Appendix C and in the file appendixch6.nb.

6.5.2 Constrained optimization with a barrier method

Our second approach to solving (Prob$_1$) is to restate it as a *constrained optimization problem* (Prob$_2$), and then solve it using a *barrier method* that we shall describe in this subsection.

We begin by recalling that (Prob$_1$) has two performance requirements: a weighted sensitivity constraint, and a closed-loop magnitude constraint. In section 6.5.1 we discussed how to solve (Prob$_1$) by optimizing a performance function defined in terms of both constraints. Now we state a problem of optimizing a performance function defined in terms of the weighted sensitivity function, over closed-loop functions T that satisfy the closed-loop magnitude constraint. This problem is equivalent to (Prob$_1$).

(Prob$_3$) Given a plant $P(s)$ and nonnegative functions $W_1(\omega)$ and $W_2(\omega)$, minimize

$$\gamma_1(T) := \sup_\omega W_1(\omega) \, | \, 1 - T(j\omega) \, |$$

subject to

$$W_2(\omega) |T(j\omega)| \leq 1$$
$$T \text{ internally stabilizes } \mathcal{S}.$$

Our practical approach to solving (Prob$_3$) is as follows. Let $\epsilon > 0$ be a fixed number, and define

$$\Gamma_\epsilon(\omega, z) := | \, W_1(\omega) \, 1 - z \, |^2 - \epsilon \log\left(1 - | \, W_2(\omega) \, z \, |^2\right) \qquad (6.44)$$

Note that Γ_ϵ is defined only for those z that satisfy

$$|W_2(\omega) z| < 1. \qquad (6.45)$$

Such z are called *feasible*. When $|W_2(\omega) z|$ gets near 1 the value of Γ_ϵ becomes quite large. Hence the logarithm in Γ_ϵ heavily punishes z for being close to violating the inequality (6.45). The logarithmic term in Γ_ϵ receives the name *barrier function*. Also, by choosing a suitable ϵ one can manipulate the contribution of the barrier function to the value of Γ_ϵ.

The algorithm for solving constrained optimization problems with barrier functions is presented now.

Optimization with the barrier method. Given $\epsilon > 0$ and a feasible function T^0,

b1. Use T^0 as initial guess to find T^* that minimizes $\sup_\omega \Gamma_\epsilon(\omega, T(j\omega))$ over all T that make $\mathcal{S}$ internally stable.

b2. Update $\epsilon \leftarrow \frac{\epsilon}{10}$, $T^0 \leftarrow T^*$.

b3. Stop if T^0 satisfies a preset tolerance criterion; else repeat (b1)–(b3).

Example. Consider $P(s) = \frac{1}{s+1}$, $W_1 = |P(j\omega)|$, and $W_2 = |P(j\omega)|^{-1}$ in (Prob$_3$). A computer run with 32 gridpoints and the barrier method initialized with $\epsilon = 1$ and $T^0(s) = 0.1 + 0.5\frac{s-1}{s+1}$ produces

$$T^b(s) = \frac{1.77066 + 2.69574s + 0.958249s^2}{3.31942 + 3.41184s + 1.0s^2} \tag{6.46}$$

and the results shown in Table 6.2. Also see Fig. 6.48.

Table 6.2. *Computer run with the barrier method.*

ϵ	Iter.	Flat	GrAlign	$\sup_\omega \|W_1(1-T)\|$	$\sup_\omega \|W_2 T\|$
1.0	5	0.0004	0	0.59	0.64
0.1	5	0.0008	0	0.46	0.96

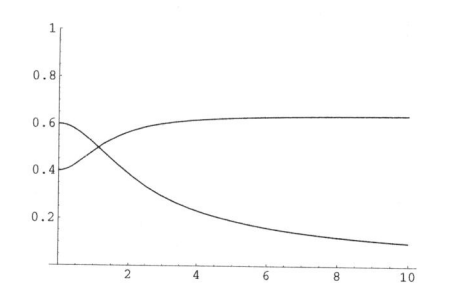

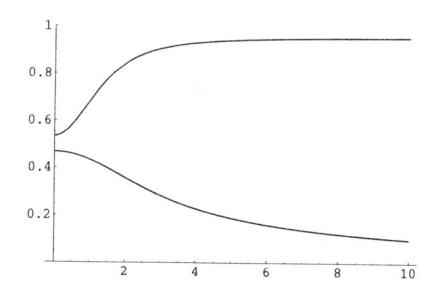

Fig. 6.48. *Plot of $W_1(\omega)|1 - T(j\omega)|$ and $W_2(\omega)|T(j\omega)|$ for the optimal T obtained with $\epsilon = 1$ and $\epsilon = 0.1$.*

Conclusion. The barrier method with 32 gridpoints was used to solve (Prob$_1$). A solution is already obtained with $\epsilon = 1$. The weighted sensitivity attains a maximum value of 0.59, and the weighted magnitude of the closed-loop function attains a maximum value of 0.64.

For comparison another solution was obtained using $\epsilon = 0.1$. In this case the weighted sensitivity attains a maximum value of 0.46 and the weighted magnitude of the closed-loop attains a maximum value of 0.96. Thus reducing ϵ leads to higher weighted magnitude of the closed loop function (but still acceptable), and leads to lower magnitude of the weighted sensitivity. A computer code template for doing runs similar to this one is given in Appendix C and in the file appendixch6.nb.

Chapter 7

Internal Stability II

Internal stability was introduced in Chapter 2. This chapter continues the discussion to obtain theorems that precisely characterize internally stable systems, either in terms of zeros and poles of the plant or in terms of a formula for the closed-loop transfer function. Section 7.1 develops the mathematical tools for interpolation with rational functions. In section 7.2 a characterization of internally stable systems is given in terms of interpolation conditions on the closed-loop transfer function, when the plant has simple RHP zeros and poles. The case of higher multiplicity is treated in section 7.3.

7.1 Calculating interpolants

Systems $\mathcal{S}$ whose plant has RHP poles or RHP zeros have closed-loop transfer functions that can be described by a formula. We shall see that this formula depends on those RHP poles and RHP zeros of the plant and nothing else. To derive the formula and related results, we concentrate first on the case where all of these zeros and poles have multiplicity 1 (called *simple* poles and zeros).

Suppose that the following sets of complex numbers are given:

$$s_1, \ldots, s_n \in \text{RHP} \quad \text{(data points)} \tag{7.1}$$

and

$$v_1, \ldots, v_n \in C \quad \text{(data values)} . \tag{7.2}$$

For $T(s)$ a rational function, consider the set of equalities

$$(INT) \qquad T(s_1) = v_1, \quad T(s_2) = v_2, \quad \ldots, \quad T(s_n) = v_n$$

called an *interpolation condition on* T. If $T \in \mathcal{RH}^\infty$ satisfies INT, then T is called an *interpolant* (for the data given in (7.1) and (7.2)).

The main objective of this section is to present ways to generate interpolants. More precisely, given a set of equalities INT we show how to

- Produce one interpolant.

- Produce a parameterization (formula) of all interpolants.

First, we present an elementary way to produce one interpolant (section 7.1.1) and to parameterize all interpolants (section 7.1.2). This is then generalized to problems with an interpolation condition at infinity (section 7.1.3).

Interpolation with polynomials, rather than rational functions, is a standard subject in numerical analysis (see [DC80]). The treatment here for rational functions is conceptually the same, and proofs are similar.

7.1.1 Calculating one interpolant

Let INT and a real number $a < 0$ be given. We will show how to produce an interpolant T in $\mathcal{RH}^\infty$ such that all its poles are located at $s = a$.

Let T be a proper rational function of s with a single pole location at $s = a$. One can think of T as being a sum of rational functions whose denominators are of the form $(s-a)^k$, with $k = 0, \ldots, n-1$. Since T is proper, for each factor $(s - s_0)$ that appears in the numerator there is a corresponding factor $(s - a)$ in the denominator. Thus T must have the form

$$T(s) = c_0 + c_1 \frac{s-s_1}{s-a} + c_2 \frac{(s-s_1)(s-s_2)}{(s-a)^2} + \cdots + c_{n-1} \frac{(s-s_1)\cdots(s-s_{n-1})}{(s-a)^{n-1}} \tag{7.3}$$

for some constants $c_0, c_1, \cdots, c_{n-1}$. The right-hand side of (7.3) is called *Newton's representation for the function $T(s)$*.

Now we seek $T \in \mathcal{RH}^\infty$ with the form (7.3) that satisfies INT. Set $s = s_1, s = s_2, \ldots, s = s_n$ in (7.3) and combine with INT to obtain the system of equations

$$\begin{aligned} c_0 &= v_1 \\ c_0 + c_1 \frac{s_2-s_1}{s_2-a} &= v_2 \\ &\vdots \\ c_0 + \cdots + c_{n-1} \frac{(s_n-s_1)\cdots(s_n-s_{n-1})}{(s_n-a)^{n-1}} &= v_n. \end{aligned} \tag{7.4}$$

This system of equations in the unknown c_k has a unique solution. It can be solved easily by back substitution.

Example. Find $T \in \mathcal{RH}^\infty$ with $a = -1$ as its only pole location, such that

$$T(0) = 1, \quad T(1) = -2. \tag{7.5}$$

Solution. Note that $n = 2$ (number of points in INT), so Newton's representation for an interpolant is

$$T(s) = c_0 + c_1 \frac{s}{s+1}. \tag{7.6}$$

Combining (7.5) and (7.6), we see that (7.3) becomes

$$\begin{aligned} c_0 &= v_1 \\ c_0 + c_1 \frac{1}{2} &= v_2 \end{aligned} \tag{7.7}$$

7.1. CALCULATING INTERPOLANTS

and we obtain $c_0 = 1$ and $c_1 = -6$.

A special situation occurs when the interpolation values $v_1, \ldots, v_n$ are all zero. In this case, system (7.4) gives $c_0 = c_1 = \ldots = c_{n-1} = 0$, so the function T given by the Newton's representation method is the zero function. A nontrivial interpolant for this case is easily obtained by setting

$$T_0(s) = \gamma \frac{(s - s_1) \cdots (s - s_n)}{(s + a)^n} \tag{7.8}$$

where γ is any nonzero real constant. Observe that the function T_0 in (7.8) has no zeros in the RHP other than $s_1, \ldots, s_n$.

7.1.2 Parameterization of all interpolants

Given a set INT of interpolation conditions, by INT_0 we denote the set of homogeneous interpolation conditions

$$(INT)_0 \qquad T(s_1) = 0, \ \ T(s_2) = 0 \ \ , \ldots, \ \ T(s_m) = 0$$

obtained from INT by setting the right-hand sides to 0. The set of conditions INT_0 plays a fundamental role in parameterizing all functions $T \in \mathcal{RH}^\infty$ that satisfy INT.

THEOREM 7.1.1. *Let a set INT of interpolation conditions be given. Also given are functions T_1, T_0 in $\mathcal{RH}^\infty$ such that T_1 satisfies INT, T_0 satisfies INT_0, $s_0, \ldots, s_n$ are the only zeros of T_0, and T_0 has relative degree 0. Then every T in $\mathcal{RH}^\infty$ that satisfies INT has the form*

$$T(s) = T_1(s) + T_0(s) F(s) \tag{7.9}$$

for some F in $\mathcal{RH}^\infty$. Conversely, if a function T is given by the formula (7.9), then T belongs to $\mathcal{RH}^\infty$ and satisfies INT.

Proof. Let T be a function in $\mathcal{RH}^\infty$ that satisfies INT. Then the difference $T(s) - T_1(s)$ has zeros at the locations $s = s_1, \ldots, s = s_n$. Therefore the rational function $F(s) = (T(s) - T_1(s))/T_0(s)$ has no poles in the closed RHP and is proper since T_0 has relative degree 0. This says that $F(s)$ belongs to $\mathcal{RH}^\infty$, and we have proved the first part of the theorem. To prove the second part, suppose that T in $\mathcal{RH}^\infty$ satisfies (7.9). By direct substitution of the values $s = s_0, s_1, \ldots, s_n$ in the formula (7.9) we see that T satisfies INT. ∎

It follows from Theorem 7.1.1 that it is enough to find particular functions T_0 and T_1, as in formula (7.9), to determine all $\mathcal{RH}^\infty$ functions that satisfy INT.

Example. We will find a formula for all functions $T \in \mathcal{RH}^\infty$ that satisfy

$$T(1) = 2, \ \ T(3) = 4 \tag{7.10}$$

We can choose any negative number as the location of the pole of the rational functions T_0 and T_1 in formula (7.9). We select $a = -2$; set

$$T_0(s) = \frac{(s-1)(s-3)}{(s+2)^2} \tag{7.11}$$

and

$$T_1(s) = c_0 + c_1 \frac{s-1}{s+2}. \tag{7.12}$$

By taking $s = 1$ and $s = 3$ in (7.12) and combining with (7.10), we obtain $c_0 = 2$ and $c_1 = 5$. We know from Theorem 7.1.1 that all functions $T \in \mathcal{RH}^\infty$ that satisfy (7.10) are of the form

$$T(s) = 2 + 5\frac{s-1}{s+1} + \frac{(s-1)(s-3)}{(s+1)^2}F(s), \tag{7.13}$$

where $F \in \mathcal{RH}^\infty$.

7.1.3 Interpolation with a relative degree condition

If a rational function T is strictly proper, then as $|s|$ gets large, $|T(s)|$ approaches 0. One can think of T as having a zero at $s = \infty$. The order we associate with this "zero" is the same number as the *relative degree of* T, $d(T)$ (the difference of degrees between denominator and numerator). One way to incorporate this special zero in interpolation problems is to specify the relative degree of the interpolant together with *INT*.

Given *INT* and an integer $r \geq 0$, we seek T in $\mathcal{RH}^\infty$ with relative degree $d(T) = r$ that satisfies *INT*. To produce one interpolant, modify the representation (7.3) to include the requirement on the relative degree of T:

$$T(s) = \frac{1}{(s-a)^r} \times \left(c_0 + c_1\frac{s-s_1}{s-a} + c_2\frac{(s-s_1)(s-s_2)}{(s-a)^2} + c_{n-1}\frac{(s-s_1)\cdots(s-s_{n-1})}{(s-a)^{n-1}}\right) \tag{7.14}$$

To find $c_0, c_1, \ldots, c_{n-1}$, proceed as before by forming a triangular system of equations and solving it.

A parameterization of all interpolants that satisfy both *INT* and the relative degree condition $d(T) = r$ is obtained after a small modification to Theorem 7.1.1. Details of the proof of Theorem 7.1.2 are left to the reader.

THEOREM 7.1.2. *Assume the hypotheses of Theorem 7.1.1, with the additional assumptions that $d(T_1) \leq r$ and $d(T_0) = r$. Then a rational function T is a function in $\mathcal{RH}^\infty$ that has relative degree r and satisfies INT if and only if there exists a rational $F \in \mathcal{RH}^\infty$ such that*

$$T(s) = T_1(s) + T_0(s)F(s) \tag{7.15}$$

7.2 Plants with simple RHP zeros and poles

Internal stability can be expressed as the condition that certain functions associated with $\mathcal{S}$ are bounded and have no poles in the closed RHP. This fact is presented below as a lemma needed for the main theorems of this section, but it is interesting in itself. In fact, it can be used as an alternative definition of internal stability that generalizes to MIMO systems (see [BB91], [DFT92], and [F87]).

LEMMA 7.2.1. *The closed-loop system of Fig. 2.1 is internally stable if and only if*

$$Q, T, S, SP \quad \text{belong to} \quad \mathcal{RH}^\infty. \tag{7.16}$$

Proof. Key relations are $S = 1 - T$ and $T = QP = PQ$. These imply that T is in $\mathcal{RH}^\infty$ if and only if S and $PQ = QP$ are in $\mathcal{RH}^\infty$. If the system is internally stable, we only have to check that $Q = CS$ and SP belong to $\mathcal{RH}^\infty$. Note that

$$S(s_p) = \frac{1}{1+PC}(s_p) = 0 \tag{7.17}$$

at the poles s_p of C. Thus $Q = CS$ is in $\mathcal{RH}^\infty$. Similar reasoning implies that

$$SP = \frac{1}{1+PC}P \tag{7.18}$$

is in $\mathcal{RH}^\infty$. This proves half of the lemma. Now suppose that the relations in (7.16) hold. If there is pole-zero cancellation in PC at a point $s = s_0$ in the RHP, then either $Q = C/(1+PC)$ or $SP = P/(1+PC)$ has a pole at $s = s_0$, both of which contradict the hypothesis. ∎

To illustrate the use of this lemma, we verify (7.16) for the system $\mathcal{S}_1$ with plant $P(s) = 1/(s-1)$ and compensator $C_1(s) = (s-1)/(s+1)$. In section 2.3 we saw that $T_1(s) = 1/(s+2)$, which belongs to $\mathcal{RH}^\infty$. We also have

$$P(s)S_1(s) = P(s)(1-T_1(s)) = \frac{1}{s-1}\left(1 - \frac{1}{s+2}\right) = \frac{s+1}{(s-1)(s+2)}. \tag{7.19}$$

We see that the closed-loop plant PS_1 has a pole at $s = 1$. By Lemma 7.2.1 we can assert that the system $\mathcal{S}_1$ is not internally stable.

Now consider the system $\mathcal{S}_2$ with the same plant and compensator $C_2(s) = 2$. Recall that $T_2(s) = 2/(s+1)$, so

$$\begin{aligned} P(s)S_2(s) &= \tfrac{1}{s-1}\tfrac{s-1}{s+1} &= \tfrac{1}{s+1} \\ Q_2(s) &= C_2(s)S_2(s) &= 2\tfrac{s-1}{s+1}. \end{aligned} \tag{7.20}$$

Therefore (7.16) is satisfied, and the system $\mathcal{S}_2$ is internally stable.

The description of internal stability in terms of interpolation conditions is given in the following theorem.

THEOREM 7.2.2. *Consider the closed-loop system $\mathcal{S}$, where the plant P is given. Let ρ_ℓ ($\ell = 1, \ldots, n$) denote the poles and z_ℓ ($\ell = 1, \ldots, m$) denote the zeros of the plant P in the closed RHP. Suppose that all unstable zeros and poles of the plant are simple. If $\mathcal{S}$ is internally stable, then there exists an integer $d_c \geq 0$ such that the closed-loop transfer function $T = PC(1 + PC)^{-1}$ must satisfy interpolation conditions*

$$\begin{cases} T(\rho_\ell) = 1 & (\ell = 1, \ldots, n) \\ T(z_\ell) = 0 & (\ell = 1, \ldots, m) \\ d(T) = d(P) + d_c. \end{cases} \tag{7.21}$$

Conversely, if T is any function in $\mathcal{RH}^\infty$ satisfying (7.21) for some $d_c \geq 0$, then the closed-loop system $\mathcal{S}$ associated with T and P is internally stable.

Proof. Suppose the system $\mathcal{S}$ is internally stable. Then $P^{-1}T = C(1 + PC)^{-1} = Q \in \mathcal{RH}^\infty$. Clearly, if $P(z_\ell) = 0$, then z_ℓ is a pole of P^{-1}. Thus $T(z_\ell) = 0$. Also $SP = (1 + PC)^{-1}P \in \mathcal{RH}^\infty$, so poles of P are zeros of $(1 + PC)^{-1} = 1 - T$. Then $T(\rho_\ell) = 1$. We now set $d_c = d(C)$ in equation (7.21) to finish the first part of the proof.

Now suppose that $T \in \mathcal{RH}^\infty$ satisfies (7.21). The relation $d(T) - d(P) = d_c \geq 0$ combined with (7.21) ensures that the compensator is proper. This fact, and both P and T having the same RHP zeros imply that $Q = TP^{-1}$ is proper and has no RHP poles, i.e., $P^{-1}T = Q \in \mathcal{RH}^\infty$. Also $S = 1 - T \in \mathcal{RH}^\infty$. Again, (7.21) says that poles of P are zeros of $1 - T$; thus $(1 - T)P = SP \in \mathcal{RH}^\infty$. Therefore $\mathcal{S}$ is internally stable. ∎

Theorem 7.2.2 can be combined with Theorem 7.1.2 to obtain a formula describing all possible internally stable systems $\mathcal{S}$ associated with a fixed plant. The proof of Theorem 7.2.3 is left to the reader.

THEOREM 7.2.3. *Let P be a proper, real, rational function, with simple RHP zeros $z_1, \ldots, z_n$ and RHP poles $\rho_1, \ldots, \rho_m$, and let d_c be a nonnegative integer. Let T_0, T_1 be $\mathcal{RH}^\infty$ functions such that*

$$\begin{cases} T_1(\rho_\ell) = 1 & (\ell = 1, \ldots, n) \\ T_1(z_\ell) = 0 & (\ell = 1, \ldots, m) \\ d(T_1) \geq d(P) + d_c \end{cases} \tag{7.22}$$

and

$$\begin{cases} T_0(\rho_\ell) = 0 & (\ell = 1, \ldots, n) \\ T_0(z_\ell) = 0 & (\ell = 1, \ldots, m) \\ d(T_0) = d(P) + d_c. \end{cases} \tag{7.23}$$

and suppose that all the RHP zeros of T_0 are listed in (7.23).

If $T \in \mathcal{RH}^\infty$ is the closed-loop transfer function of an internally stable system $\mathcal{S}$ with plant P and $d(C) = d_c$, then there exist $H \in \mathcal{RH}^\infty$ such that

$$T(s) = T_1(s) + T_0(s)H(s). \tag{7.24}$$

7.3. PARAMETERIZATION: THE GENERAL CASE

Conversely, if (7.24) holds for some $H \in \mathcal{RH}^\infty$, then the system $\mathcal{S}$ associated with P and T is internally stable, and the degree of its compensator is $d(C) = d_c$.

Example. Consider the plant $P(s) = (s-1)/(s-3)$, and set $d_c = 0$. We now find a formula for all T that yields an internally stable system $\mathcal{S}$ with $d(C) = 0$ and plant $P(s) = (s-1)/(s-3)$.

The plant P has RHP zero $s = 1$ and RHP pole $s = 3$. Set $T_1(s) = 2(s-1)/(s+1)$ and $T_0(s) = (s-1)(s-3)/(s+1)^2$. Then T_1 and T_0 satisfy (7.22) and (7.23), respectively, and by Theorem 7.2.3 we have

$$T(s) = \frac{2(s-1)}{s+1} + \frac{(s-1)(s-3)}{(s+1)^2} H(s) \quad (H \in \mathcal{RH}^\infty).$$

7.3 Parameterization: The general case

In this section, the results already obtained on parameterization of internally stable systems is generalized to the case where the plant may have RHP zeros or poles with multiplicity greater than 1. Results in this section are stated without proof.

7.3.1 Higher-order interpolation

The interpolation conditions we now consider have specific values of the derivatives of the function at selected points.

$$\mathbf{INT}_h \begin{cases} T(s_1) = v_{1,1}, & T^{(1)}(s_1) = v_{1,2} & \ldots, & T^{(m_1-1)}(s_1) = v_{1,m_1} \\ T(s_2) = v_{2,1}, & T^{(1)}(s_2) = v_{2,2} & \ldots, & T^{(m_2-1)}(s_2) = v_{2,m_2} \\ \vdots & \vdots & & \vdots \\ T(s_n) = v_{n,1}, & T^{(1)}(s_n) = v_{n,2} & \ldots, & T^{(m_n-1)}(s_n) = v_{n,m_n} \\ d(T) = r, \end{cases}$$

where $r \geq 0$, $\{s_1, s_2, \ldots, s_n\} \subset \text{RHP}$,[1] and

$$\{v_{11}, \ldots, v_{1m_1}, v_{21}, \ldots, v_{2m_2}, \ldots, v_{n1}, \ldots, v_{nm_n}\} \subset \mathbb{C}$$

are given. As in previous sections, we consider the problems of producing one interpolant T_1 for INT_h and of obtaining a parameterization of all interpolants for INT_h.

We use two $\mathcal{RH}^\infty$ functions, T_0 and T_1. The assumptions on these are that T_1 satisfies INT_h and that T_0 satisfies the corresponding set of homogeneous

---

[1] The inclusion of the condition $\{s_1, \ldots, s_n\} \subset \text{RHP}$ is not required for interpolation. However, it is essential for internal stability.

conditions

$$\mathbf{INT}_h^0 \begin{cases} T(s_1) = 0, & T^{(1)}(s_1) = 0 & \ldots, & T^{(m_1-1)}(s_1) = 0 \\ T(s_2) = 0, & T^{(1)}(s_2) = 0 & \ldots, & T^{(m_2-1)}(s_2) = 0 \\ \vdots & \vdots & & \vdots \\ T(s_n) = 0, & T^{(1)}(s_n) = 0 & \ldots, & T^{(m_n-1)}(s_n) = 0 \\ d(T) = r. \end{cases}$$

Furthermore, we suppose that all the RHP zeros of T_0 are listed in INT_h^0 and that $T_0^{(m_\ell)}(s_\ell) \neq 0$ ($\ell = 1, \ldots, n$).

A variation of the method we have been using works for higher-order interpolation problems as well. The difference here is that the formula for T now contains summands with numerators of the form 1, $(s - s_1)$, $(s - s_1)^2$, ..., $(s - s_1)^{m_1-1}$, $(s - s_2)(s - s_1)^{m_2-1}$, ...

THEOREM 7.3.1. *A function T is in $\mathcal{RH}^\infty$, has relative degree r, and satisfies INT_h if and only if there exists $H \in \mathcal{RH}^\infty$ such that*

$$T(s) = T_1(s) + T_0(s)H(s). \tag{7.25}$$

Example. Find $T \in \mathcal{RH}^\infty$ with relative degree $d(T) = 0$ such that

$$T(0) = 1, \ T^{(1)}(0) = 2, \ T(3) = 4, \ T(4) = 0. \tag{7.26}$$

Solution. The expression for T is

$$T(s) = c_0 + c_1 \frac{s}{s+1} + c_2 \frac{s^2}{(s+1)^2} + c_3 \frac{s^2(s-3)}{(s+1)^3}. \tag{7.27}$$

We also need the derivative of T,

$$T^{(1)}(s) = \frac{c_1}{(s+1)^2} + \frac{2c_2 s}{(s+1)^3} + \frac{c_3(-6s + 6s^2)}{(s+1)^4}. \tag{7.28}$$

Setting $s = 0$ in (7.27) and in (7.28) gives $c_0 = 1$ and $c_1 = 2$. Now using $s = 3$ in (7.27) gives

$$4 = c_0 + c_1 \frac{3}{4} + c_2 \frac{9}{16} = 1 + 2 \frac{3}{4} + c_2 \frac{9}{16}. \tag{7.29}$$

Solving for c_2 in (7.29) yields $c_2 = \frac{8}{3}$. Finally setting $s = 4$ in (7.27) leads to

$$0 = c_0 + c_1 \frac{4}{5} + c_2 \frac{16}{25} + c_3 \frac{32}{125}, \tag{7.30}$$

and this implies that $c_3 = \frac{-1615}{48}$.

7.3.2 Plants with high-multiplicity RHP zeros and poles

A RHP pole or zero of the plant with multiplicity higher than 1 in an internally stable system can be translated as an interpolation condition on derivatives of T at that location. More precisely:

THEOREM 7.3.2. *Consider the closed-loop system* $\mathcal{S}$. *Let* ρ_ℓ ($\ell = 1, \ldots, n$) *denote the poles and* z_ℓ ($\ell = 1, \ldots, m$) *denote the zeros of the plant* P *in the closed* RHP, *so that these poles and zeros have multiplicity* n_ℓ, ($\ell = 1, \ldots, n$) *and* m_ℓ ($\ell = 1, \ldots, m$), *respectively. If* $\mathcal{S}$ *is internally stable, then for some integer* $d_c \geq 0$ *the closed-loop transfer function* $T = PC(1+PC)^{-1}$ *satisfies the interpolation conditions*

$$INT_h \begin{cases} T(\rho_\ell) = 1, & T^{(1)}(\rho_\ell) = 0 \quad \ldots, \quad T^{(n_\ell - 1)}(\rho_\ell) = 0 \quad (\ell = 1, \ldots, n) \\ T(z_\ell) = 0, & T^{(1)}(z_\ell) = 0 \quad \ldots, \quad T^{(m_\ell - 1)}(z_\ell) = 0 \quad (\ell = 1, \ldots, m) \\ d(T) = d(P) + d_c. \end{cases}$$

(7.31)

Conversely, if T *is any function in* $\mathcal{RH}^\infty$ *that satisfies* INT_h *for some* $d_c \geq 0$, *then the closed-loop system* $\mathcal{S}$ *associated with* T *is internally stable and its compensator has relative degree* d_c.

We now state the result with the parameterization of internally stable systems.

THEOREM 7.3.3. *Assume the hypotheses of Theorem 7.3.2, and let* T_1, T_0 *in* $\mathcal{RH}^\infty$ *be such that* T_1 *satisfies* INT_h *and* T_0 *satisfies* INT_h^0. *Furthermore, suppose that all the RHP zeros of* T_0 *are listed in* INT_h^0 *and that* $T_0^{n_\ell}(\rho_\ell) \neq 0$ ($\ell = 1, \ldots, n$), $T_0^{m_\ell}(z_\ell) \neq 0$ ($\ell = 1, \ldots, m$).

If $T \in \mathcal{RH}^\infty$ *is the closed-loop transfer function of an internally stable system* $\mathcal{S}$ *with plant* P *and* $d(C) = d_c$, *then there exist* $H \in \mathcal{RH}^\infty$ *such that*

$$T(s) = T_1(s) + T_0(s)H(s). \tag{7.32}$$

Conversely, if (7.32) holds for some $H \in \mathcal{RH}^\infty$, *then the system* $\mathcal{S}$ *associated with* P *and* T *is internally stable, and the degree of its compensator is* $d(C) = d_c$.

Example. If $P(s) = \frac{(s+3)(s-2)}{s^2}$ and $d_c = 1$, then the interpolation conditions on any $T \in \mathcal{RH}^\infty$ for internal stability of the system $\mathcal{S}$ are given by

(INT) $\quad T(0) = 1, \quad T'(0) = 0, \quad T(2) = 0, \quad d(T) = 1.$

If $a = -1$ is selected as the location of the poles of T, then the technique described in section 7.3.1 yields the interpolant

$$\frac{-15s^2}{4(1+s)^3} + \frac{s}{(1+s)^2} + \frac{1}{1+s}. \tag{7.33}$$

Thus we have the parameterization

$$T(s) = \frac{-15s^2}{4(1+s)^3} + \frac{s}{(1+s)^2} + \frac{1}{1+s} + \frac{s^2(s-2)}{(1+s)^4} T_1(s) \quad \forall T_1 \in \mathcal{RH}^\infty. \tag{7.34}$$

7.4 Exercises

1. Find an interpolant for each set of conditions below.

    a. $T(1) = 0$, pole location $s = -1$
    b. $T(2+j) = 3$, $T(2-j) - 3$, pole location $s = -10$
    c. $T(0) = 0$, $T(1) = 0$, $T(2) = 0$, pole location $s = -5$
    d. $T(1) = 2$, $T(3+j) = 0$, $T(3-j) = 0$, pole location $s = -1$

2. Prove that the system of equations (7.4) does have a solution in $c_0, \ldots, c_{n-1}$ and that this solution is unique.

3. Prove that if the function $T \in \mathcal{RH}^\infty$ satisfies $T(s_1) = 0, \ldots, T(s_n) = 0$, then either T is identically zero or degree{denominator(T)} $\geq n$.

4. Prove that given INT with n data points, *there exists a unique* $T \in \mathcal{RH}^\infty$ that satisfies INT, with the following property: if p and q are polynomials such that $T = p/q$, then degree$(q) < n$.

5. Find *all* interpolants satisfying the conditions stated in Exercises 1a–d.

6. Prove Theorem 7.1.1.

7. Find an interpolant for each set of conditions below.

    a. $T(1) = 0$, pole location $s = -1$, relative degree $d(T) = 2$
    b. $T(2+j) = 3$, $T(2-j) = 3$, pole location $s = -10$, $d(T) = 1$
    c. $T(0) = 2$, $T(j) = 1 + 2j$, $T(-j) = 1 - 2j$, pole location $s = -1$, $d(T) = 1$

8. Prove the statements below.

    a. If both $T_1, T_2 \in \mathcal{RH}^\infty$ have relative degree r, then $T_1 + T_2$ has relative degree $\leq r$. Give an example where strict inequality occurs.
    b. If $T_1, T_2 \in \mathcal{RH}^\infty$ have relative degrees r_1, r_2, respectively, then the product $T_1 T_2$ has relative degree $r_1 + r_2$.

9. Find *all* functions $T \in \mathcal{RH}^\infty$ that satisfy the conditions stated in Exercise 7a – 7c.

10. Prove Theorem 7.1.2.

11. Find a formula for all closed-loop transfer functions T that correspond to an internally stable system $\mathcal{S}$, if the relative degree of the compensator is 0 and the plant $P(s)$ is given by

    a. $1/(s-3)$
    b. $(s^2 - 4)/(s^2 - 1)$

7.4. EXERCISES

      c. $(s-5)/(s^2+1)$

      d. $(s+2)/(s+4)$

12. Do Exercise 11a–11d, but with $d(C) = 1$.

13. Use the results in section 7.3 to find the closed-loop T that arise from internally stable systems with degree of compensator $d_c = 0$ and plant

      a. $(s^2 - 2s + 1)/(s^2 - 4)$

      b. $1/(s-2)^2$

      c. $(s+2)/(s-2)^2$

      d. Do problems a–c with $d(C) = 1$.

14. Supply a proof for the results stated in section 7.3.

References and further reading

References that amplify our presentation of internal stability, interpolation, and control (but with a different point of view) are [BGR90], [BB91] (gives a good history), [DC80], [F87], [DFT92], [V85], [YJB76a], and [YJB76b].

Part III
H^∞ Theory

Appendix A

History and Perspective

While historians trace control to Archimedes' time and the theory of control to a paper of James Clerk Maxwell on governors, the subject called classical control began in the Second World War in labs in England and the United States that designed radar-driven anti-aircraft guns. The beginnings of classical control are summarized in a book [JNP47] by James, Nichols, and Phillips. According to Ralph Phillips, a book that was very influential in their lab was Bode's famous book on amplifiers, although Bode's book is not referenced in [JNP47]. The primary technique that emerged was adjusting parameters in low order (e.g., degree 2) rational functions and checking that the graphs lie in certain regions. Classical control dominated industrial practice for many years, even though it could be taught only by example and its main technique was trial and error.

Much of the theory of control in the 1960s and 1970s focused on achieving desired frequency domain performance as closely as possible in a mean-square-error sense. We have nothing to add to the literature here and so do not give a historical treatment.

The subject of optimizing worst-case error in the frequency domain along its present lines started not with control but with circuits. One issue was to design amplifiers with maximum gain over a given frequency band. Another was the design of circuits with minimum broadband power loss. Indeed, H^∞ control is a subset of a broader subject, H^∞ *engineering*, which focuses on worst-case design in the frequency domain. In paradigm engineering problems this produces what the mathematician calls an "interpolation problem" for analytic functions. The techniques of Nevanlinna–Pick interpolation had their first serious introduction into engineering in a SISO circuits paper by Youla and Saito [YS67] in the mid-1960s. Further development waited until the mid-seventies, when Helton [H76], [H78], [H81] applied interpolation and more general techniques from operator theory to amplifier problems. Here the methods of commutant lifting [A63], [NF70], [S67] and of Admajan–Arov–Krein (AAK) [AAK68], [AAK72], [AAK78] were used to solve MIMO optimization problems. The disk method used in this context was first described in an engineering article on gain equalization [H81] and followed the foundational theory published in the mathematics article [H78] three years earlier.

In the late 1970s G. Zames [Z79] began to marshal arguments indicating that H^∞ rather than H^2 was the physically proper setting for control. Zames suggested on several occasions that these methods were the appropriate ones for codifying classical control. These efforts yielded a mathematical problem that Helton identified as an interpolation problem solvable by existing means (see [ZF81]). In 1981 Zames and Francis [ZF83] used this to solve the resulting single-input, single-output (SISO) problem. In 1982 Chang–Pearson [CP84] and Francis–Helton–Zames [FHZ84] solved it for a many-input, many-output (MIMO) system.

The pioneering work of Zames and Francis treated only sensitivity optimization. In 1983 three independent efforts emphasized bandwidth constraints, formulated the problem as a precise mathematics problem, and indicated effective numerical methods for its solution: Doyle [D83], Helton [H83], and Kwakernaak [K83]. Helton treated the problem graphically, just as it is done in Part I of this book, by approximating multiple-disk constraints with a single moving disk. Kwakernaak gives an equivalent method, but it is expressed algebraically in terms of two weight functions W_1 and W_2 on the sensitivity function $I - T$ and on T. Also, all of these papers described quantitative methods which were soon implemented on computers. It was these papers that actually laid out precisely the trade-off in control between performance at low frequency and roll-off at higher frequency and how one solves the resulting mathematics problem. This is in perfect analogy with amplifier design where one wants large gain over as wide a band as possible, producing the famous gain-bandwidth trade-off. Rather remarkable in textbook treatments of control through the 1970s was the absence (or cursory treatment) of bandwidth constraints. On the other hand, many control practitioners of that period would say that bandwidth constraints were important, but when they went to formulate them as a mathematical optimization problem they did it by dropping all high-frequency constraints. Thus they made what the authors like to call the fundamental mistake of H^∞ control (described in section 10.4 in [HMer98]), and the concomitant confusion cost the methods much credibility. As of the 1990s the control community's views have become more realistic. This conceptual shift is one of the main accomplishments of H^∞ control. The moral of the story is that nothing stabilizes a nebulous philosophical discussion like the ability to solve the math problems that arise. Mush dries up quickly under this torch.

All of the traditional methods in H^∞ solved only optimization problems where the specification sets S_θ are disks. There is a beautiful and extensive theory of this. Much of the mathematical theory came directly from existing work in the study of operators on Hilbert space. However, an important accomplishment of theoretical engineers was to formulate the problem in a new set of coordinates (called state-space coordinates) that are well suited to many problems. This extremely elegant theory is described in [BGR90], [F87], [GM90], [Gl84], [ZDG96].

To describe the origins of state-space H^∞ engineering we must back up a bit. Once the power of the commutant lifting – AAK techniques were demonstrated

APPENDIX A. HISTORY AND PERSPECTIVE

on engineering problems, P. de Wilde played a valuable role by introducing them to signal processing applications (see [deWVK78]) and to others in engineering. The state-space solutions of H^∞ optimization problems originated not in H^∞ control, but in the area of model reduction. The AAK work with a shift of language is a paper on model reduction (though not in state-space coordinates) by Bettayab–Safanov–Silverman [BSS80], which gives a statespace viewpoint for SISO systems. Subsequently Glover [Gl84] gave the MIMO state-space theory of AAK-type model reduction. Since the H^∞ control problem was already known to be solvable by AAK, this quickly gave statespace solutions to the H^∞ control problem. These state-space solutions were described first in 1984 by Doyle [D_{report}], which though never published was extremely influential. Earlier, in his thesis (unpublished), he had given state-space H^∞ solutions based on converting the geometric (now called behavioral by engineers) version of commutant lifting–AAK due to Ball and Helton to state-space.

As basic as it is, the work on internal stability treated in Part I was done fairly recently. In the famous article of Youla–Jabr–Bongiorno [YJB76a], [YJB76b], the concept of internal stability was introduced and reduced to a mathematical problem in which all H^∞ functions meet a given interpolation constraint. The authors then solved the H^2 control problem (rather than optimizing over H^∞, as is done in this book).

Independent of the rise of H^∞ control was the Horowitz approach [Ho63]. The relationship between his approach and ours is described in [BHMer94]. Another independent development was Tannenbaum's [T80] very clever use of Nevanlinna–Pick interpolation in a control problem in 1980. Also appearing early on the H^∞ stage was Kwakernaak's polynomial theory [K86]. Another major development that dovetailed closely with the invention of H^∞ control was a tractable theory of plant uncertainty. A good historical treatment appears in [DFT92]. Another application of these techniques is to robust stabilization of systems [Kim84].

The mathematical theory behind this book is described in [HMer98] and is different from and more general than the classical mathematical theory (Nevanlinna–Pick, Nehari), which applies only to special H^∞ problems. Work on the theory for solving general H^∞ optimization problems started in 1981 by Helton and Howe, who worked on convex problems [HH86]. Since then theoretical work has been done by J. W. Helton, S. Hui, D. Marshall, O. Merino, Z. Slodkowski, A. Vityaev, K. Lenz, and E. Wegert in papers listed in the bibliography. Theoretical work pertaining to numerical issues has been continued by Helton, Merino, and T. Walker [HMer93a], [HMer93b], [HMW93]. Numerical testing and more theoretical work has been done by J. Bence, J. W. Helton, O. Merino, J. Myers, D. Schwartz, and T. Walker. Interestingly, the theory developed to solve the engineering problem *OPT* has strong ties with ongoing work in the study of functions of several complex variables. For some discussion of this, see the articles [HMer91] and [HV97]. Theorem 13.3 in [HMer98], which underlies much of the numerics employed here, is taken from [HMer93a]. Earlier versions are in the paper [H86] and independently in the pure mathematics literature a special case concerning "Kobayashi extremals" is due to Lempert [Le86]. Also interesting is recent work of Zames and Owens [OZ93] on convex H^∞ optimization. While [HH86] gave only qualitative properties of optima such as flatness, a numerical approach was proposed in [OZ95].

Some of the more recent numerical methods the authors are developing (see Chapter 19 in [HMer98]) are related to some very powerful optimization algorithms which go under the heading semidefinite programming. Modern interest in them dates to the work of Karmarkar in the mid-1980s, and a rapid evolution brought them to the form one sees now. Good references on the history of these and other methods are [Wr97], [VB96], [LO96]; we refer the reader to them.

Appendix B

Downloading OPTDesign and *Anopt*

If you are using a web browser, go to *http://anopt.ucsd.edu* and follow the directions on the web page.

Those who like doing things the hard way can download the packages OPTDesign and *Anopt* through anonymous ftp.

Type

```
ftp anopt.ucsd.edu
```

When the remote system requests the accountname, you reply

```
anonymous
```

When the system requests the password, type your email address:

```
myadress.edu
```

Then type

```
cd pub/anopt
```

Now you are in the correct directory. There are two types of files, *.tar.gz* (for Unix) and *.zip* (for MSWindows). Pick the file of your favorite type with the latest date, download it to your system, and uncompress it.

Appendix C

Computer Code for Examples in Chapter 6

This appendix contains computer code of the two sessions with the package OPTDesign, which were discussed in Chapter 6. The code can be easily modified to treat other design problems. See the notebook appendixch6.nb.

C.1   Computer code for design example 1

Calculation of the plant. The general (uncertain) plant $p_\alpha(s)$ is calculated directly from the matrices A, b, c, and d as plant$(s,\alpha) = c(sI - A)^{-1}b$. The nominal plant is plant$(s, 5)$.

```
<<OPTDesign';

A   = {{-2,  -0.4,    0.,      0.},
       {1.,-0.01,   -5.,       0.},
       {0.,  10.0,    0.,     -10},
       {0.,   0.0,alpha,-alpha}};
b   = {2,0,0,0};
c   = {0,1,0,0};
d   = {0};

plant[alpha_,s_]  = c.Inverse[ s IdentityMatrix[4] - A].b;
p[s_]     = plant[5,s];
plantplot = BodeMagnitude[p[s]];
paux[s_]  =  1/(s+1)^2;
```

Radius and center for original list of requirements

```
r01[w_]=Which[0.0 <= Abs[w] <= 0.12,   0.1,
```

122 APPENDIX C. COMPUTER CODE FOR EXAMPLE IN CHAPTER 6

```
                  0.1 <  Abs[w] <= 1.0,    0.32,
                  1.0 <  Abs[w] <= 5.0,    2.0,
                  5.0 <  Abs[w] <= 10.0,   0.1,
                 10.0 <  Abs[w]       , (0.01/Abs[paux[I 10.]])*Abs[paux[I w]]];


k01[w_]=Which[0. <= Abs[w] <= 5.0, 1.
              5. <  Abs[s]         , 0.];

FigEnvelopePlot2D0 = EnvelopePlot[Radius->r01,Center->k01,
                                  FrequencyBand->{0.01,12}];

FigEnvelopePlot3D0 = EnvelopePlot3D[Radius->r01,Center->k01,
                                    FrequencyBand->{0.01,6}];
```

Modified center and radius functions. Lines are used to interpolate certain points chosen beforehand. In particular, the radius is set to decrease linearly from 1 to 0 between frequencies $\omega = 1$ and $\omega = 5$. The function InterpolatingPolynomial defines a line when only two interpolation points are specified.

```
line0[w_] = InterpolatingPolynomial[{{1.,1},{5.,0}},Abs[w]];

k1[w_]=Which[ 0.0 <= Abs[w] <=  1.0,1.,
              1.0 <  Abs[w] <=  5.0,line0[w],
              5.0 <  Abs[w]        ,0.];

ra[w_]=InterpolatingPolynomial[{{0.0,0.1},{0.12,0.1}}, Abs[w]];
rb[w_]=InterpolatingPolynomial[{{0.12,0.1},{1,0.32}}, Abs[w]];
rc[w_]=InterpolatingPolynomial[{{1.0,0.32},{2.0,2.0}}, Abs[w]];
rd[w_]=InterpolatingPolynomial[{{2.0,2.0},{3.0,2.0}}, Abs[w]];
re[w_]=InterpolatingPolynomial[{{3.0,2.0},{5,0.1}}, Abs[w]];
rf[w_]=InterpolatingPolynomial[{{5.0,0.1},{10.0,0.01}}, Abs[w]];


r1[w_]=Which[   0.0 <= Abs[w] <= 0.1,   ra[w],
                0.1 <  Abs[w] <= 1.0,   rb[w],
                1.0 <  Abs[w] <= 2.0,   rc[w],
                2.0 <  Abs[w] <= 3.0,   rd[w],
                3.0 <  Abs[w] <= 5.0,   re[w],
                5.0 <  Abs[w] <= 10.0,  rf[w],
               10.0 <  Abs[w], (0.01/Abs[paux[I 10.]])*Abs[paux[I w]]];

FigCenter = Show[Plot[k1[w],{w,0,7}],Plot[k01[w],{w,0,7},
             PlotStyle -> {{Thickness[0.02],GrayLevel[0.5]}}]];
plotr1a = Show[Plot[r1[w],{w,0,1}],Plot[r01[w],{w,0,1},
             PlotStyle -> {{Thickness[0.02],GrayLevel[0.5]}}]];
plotr1b = Show[Plot[r1[w],{w,0,7}],Plot[r01[w],{w,0,7},
             PlotStyle -> {{Thickness[0.02],GrayLevel[0.5]}}]];
plotr1c = Show[Plot[r1[w],{w,5,12}],Plot[r01[w],{w,5,12},
```

C.1. COMPUTER CODE FOR DESIGN EXAMPLE 1   123

```
                    PlotStyle -> {{Thickness[0.02],GrayLevel[0.5]}}]];
FigRadius = Show[GraphicsArray[{plotr1a,plotr1b,plotr1c}]];
FigEnvelopePlot2D =
        EnvelopePlot[Radius->r1,Center->k1,FrequencyBand->{0.01,12}];
FigEnvelopePlot3D =
        EnvelopePlot3D[Radius->r1,Center->k1,FrequencyBand->{0.01,6}];
```

First optimization runs and plots

```
OPTDesign[p,Center->k1,Radius->r1,Ngrid -> 256,Nsmth -> 100];

FigBodeMagS = BodeMagnitude[1-T,PlotRange -> {-30,10}];
FigBodeMagT = BodeMagnitude[T,PlotRange -> {-40,10}];
FigBodePhaT = BodePhase[T];
```

Redefinition of the center and radius functions, and plots

```
rb2[w_]=InterpolatingPolynomial[{{0.12,0.1},{1.3,0.25}}, Abs[w]];
rc2[w_]=InterpolatingPolynomial[{{1.3,0.25},{2.0,2.0}}, Abs[w]];
r2[w_] =Which[0.0 <= Abs[w] <= 0.1,    ra[w],
              0.1 <  Abs[w] <= 1.3,    rb2[w],
              1.3 <  Abs[w] <= 2.0,    rc2[w],
              2.0 <  Abs[w] <= 3.0,    rd[w],
              3.0 <  Abs[w] <= 5.0,    re[w],
              5.0 <  Abs[w] <= 10.0,   rf[w],
              10.0 < Abs[w]         , (0.01/Abs[paux[I 10.]])*Abs[paux[I w]]];

k2[w_]= k1[w];

FigCenter2 = Show[Plot[k2[w],{w,0,7}],[Plot[k01[w],{w,0,7},
                PlotStyle -> {{Thickness[0.02],GrayLevel[0.5]}}]];
FigRadius2 = Show[Plot[r2[w],{w,0,7}],Plot[r01[w],{w,0,7},
                PlotStyle -> {{Thickness[0.02],GrayLevel[0.5]}}]];

FigEnvelope2Plot2D =
        EnvelopePlot[Radius->r2,Center->k2,FrequencyBand->{0.,3}];
FigEnvelope2Plot3D =
        EnvelopePlot3D[Radius->r2,Center->k2,FrequencyBand->{0.0,3}];
```

Second optimization run and rational approximation

```
OPTDesign[p,Center->k2,Radius->r2,Ngrid -> 256,Nsmth -> 100];

TratLow[s_] = RationalModel[s,DegreeOfDenominator -> 6];
CratLow[s_] = Together[ 1/p[s] TratLow[s]/(1-TratLow[s])];

FigBodeMagTLow    = BodeMagnitude[TratLow[s],PlotRange -> {-60,10}];
FigBodePhaTLow    = BodePhase[Sample[TratLow[s]]];
```

124 APPENDIX C. COMPUTER CODE FOR EXAMPLE IN CHAPTER 6

```
FigBodeMagSLow    = BodeMagnitude[1-TratLow[s],PlotRange -> {-30,10}];
FigBodeMagCompLow = BodeMagnitude[CratLow[s]];
FigBodePhaCompLow = BodePhase[Sample[CratLow[s]]];

FigTLowZP = PlotZP[TratLow[s],s];
FigCLowZP = PlotZP[CratLow[s],s];
```

C.2 Computer code for design example 2

Setup of the problem

```
<<OPTDesign';

p[s_]    = 0.036 (s + 25)/(s^2 (s + 0.02 + I)(s + 0.02 - I));
pinv[s_] = 1/p[s];

wp = 0.7;    alphap = 0.9;
wb = 2.0;    alphab = .75;
wr = 10.;    alphar = 0.25/Abs[p[I wr]];

line1[w_] = InterpolatingPolynomial[
            {{wp,alphap Abs[pinv[I wp]]}, {wb,alphab}},w];

line2[w_] = InterpolatingPolynomial[
            {{wb,alphab},{wr,alphar Abs[p[I wr]]}},w];

line3[w_] = InterpolatingPolynomial[{{wp,1},{wb,0}},w];

k[w_]    = Which[ 0  <=  Abs[w] <= wp , 1,
                  wp <   Abs[w] <  wb , line3[Abs[w]],
                  wb <   Abs[s]       , 0];

r[w_]    = Which[0 <= Abs[w] <= wp , alphap Abs[pinv[I w]],
                 wp < Abs[w] <= wb , line1[Abs[w]],
                 wb < Abs[w] <= wr , line2[Abs[w]],
                 wr < Abs[w]       , alphar Abs[p[I w]]];

EnvelopePlot[Radius->r,Center->k,FrequencyBand->{0.01,12}];

EnvelopePlot3D[Radius->r,Center->k,FrequencyBand->{0.01,12},PlotRange->All];
```

Optimization run and model reduction

```
OPTDesign[p,Center->k,Radius->r,Ngrid->256 , Nsmth -> 30];

FigBodeMag = BodeMagnitude[T];
FigBodePha = BodePhase[T];
```

C.2. COMPUTER CODE FOR DESIGN EXAMPLE 2

```
Trat[s_]      = RationalModel[s,DegreeOfDenominator -> 7];
step1[t_]     = Chop[Simplify[InverseLaplaceTransform[Trat[s]/s,s,t]]];

FigStepT1a    = Plot[ step1[t] ,{t,0,14},PlotRange -> All];
FigStepT1b    = Plot[ step1[t] ,{t,0,14},PlotRange -> {0.9,1.2}];

Crat1[s_]     = CancelZP[ pinv[s] Trat[s]/(1-Trat[s]),s,s];
FigCrat1ZPa   = PlotZP[Crat1[s],s];
FigCrat1ZPb   = Show[FigCrat1ZPa,PlotRange -> {{-3,3},{-3,3}}];

{num,den}     = {Numerator[Crat1[s]],Denominator[Crat1[s]]};
zeros         = s /. Solve[num==0,s];
poles         = s /. Solve[den==0,s];
Crat2[s_]     = Crat1[0] *
                (1 - s/zeros[[4]])(1 - s/zeros[[5]])(1 - s/zeros[[6]])/
                 (  (1 - s/poles[[1]])(1 - s/poles[[2]])(1 - s/poles[[3]]));
FigCrat1Crat2 = Plot[{Abs[Crat1[I w]],Abs[Crat2[I w]]},{w,0,2}];
Trat2[s_]     = Together[ p[s] Crat2[s]/(1 + p[s] Crat2[s])] //Chop ;
FigCrat2ZP    = PlotZP[Crat2[s],s];
step2[t_]     = Chop[Simplify[InverseLaplaceTransform[ Trat2[s]/s,s,t] ]];
FigStepT2     = Plot[step2[t],{t,0,14}];
```

Appendix D

Anopt Notebook

Anopt: A program for sup norm optimization over spaces of analytic functions

J.W. HELTON, O. MERINO, J. MEYERS, AND T. WALKER

Lab. for Mathematics and Statistic, University of California, San Diego

D.1 Foreword

Anopt is a Mathematica package for solving diverse optimization problems over spaces of functions analytic on the unit disk in the complex plane.

The software is useful for engineers doing worst-case design in the frequency domain. This includes problems in control as well as a broadband gain equalization and matching. Another application is in the field of several complex variables where the program can be used to find analytic disks that are optimal with respect to various criteria, e.g., Kobayashi metric calculations.

The main program *Anopt[]* is easy to use, even for those with little or no computing experience. *Anopt[]* can be run at a very simple level or at a very complex level.

The package *Anopt* was developed at the Laboratory for Mathematics and Statistics at the University of California at San Diego, during the years 1989 to 1994, by J.W. Helton, Orlando Merino, Julia Myers, and Trent Walker. Financial support came from the Air Force Office of Scientific Research, the National Science Foundation, and the NSF-REU program at the San Diego Super Computer Center.

Send comments, questions, and information on bugs to anopt@math.uscd.edu.

D.2 Optimizing in the sup norm: The problem *OPT*

The mathematica program *Anopt* can be used to solve min-max problems over functions f analytic on the unit disk in the complex plane. *Anopt* can do it using several algorithms, but the default is a steepest descent–type of algorithm called disk iteration. We begin by stating the main optimization problem to be solved.

We shall use the following notation.

A_N = the space of C^N-valued analytic functions on the unit disk D in C.
$\quad f = (f[1], f[2], \ldots, f[N])$ which extend continuously to the closed disk.

$\Gamma(e, z)$ = a smooth, positive-valued function of e in the unit circle.

$$z = (z[1], z[2], \ldots, z[N]) \text{ in } C^N.$$

The main optimization problem is

$\quad$ *OPT* $\qquad$ Given $\Gamma(e, z)$, find f^* in A_N such that
$\qquad\qquad\qquad S^* = \inf_f \sup_e \Gamma(e, f(e)) = \sup_e \Gamma(e, f^*(e)).$

where the infimum is over f in A_N and the supremum is over e in the unit circle.

D.3 Example 1: First run

The problem. Our introductory example consists of a performance function where there is a single complex variable. Consider the function $\Gamma(e, z) = \text{Abs}[0.8 + 1/e + z)]^2$ as the objective or performance function. We want to solve *OPT* for this function $G(e, z)$ and calculate the optimal value and optimal function f^*.

Invoking *Anopt[]* the first time. Begin by entering Mathematica and then loading the package *Anopt*.

```
<< Anopt.m
```

$\qquad$ Anopt 3.0 ©Copyright 1991–97
$\qquad$ J.W. Helton and O. Merino. All rights reserved.

To type in the formula for the performance $\Gamma(e, z)$ requires a translation to the symbols that *Anopt* understands. To enter G in Mathematica, replace z by $z[1]$ and assign the result to a name, say g:

```
g = Abs [ 0.8 + (1/e + z[1] )^2 ]^2
```

```
             1       2  2
Abs[0.8 + (- + z[1]) ]
             e
```

D.3. EXAMPLE 1: FIRST RUN

Besides the performance function, *Anopt[]* needs as input an error tolerance. Below we set the tolerance as 0.02.

```
Anopt[g, 0.02]
```

```
It  :   Current Value      : Step    : Optimality Tests  : Error   : Sm. : Grid
    :     gammastar        :         :   flat    grAlign :   ned   :     :
---------------------------------------------------------------------------------
 0  : 3.24            E+00 :   N/A   : 9.9E-01 : 0 E+00  :   N/A   : NON : 32
 1  : 1.3456          E+00 : 4. E-01 : 4.8E-01 : 0 E+00  : 5.2E-04 : NON : 32
 2  : 1.0242720431082E+00  : 1.4E-01 : 4.9E-02 : 0 E+00  : 2.4E-05 : NON : 32
 3  : 1.0100848818098E+00  : 1.8E-02 : 1.8E-02 : 0 E+00  : 5.4E-03 : NON : 32


Summary
-------
gammastar = 1.010084881809756E+00
flat      = 1.7866711005E-02
grAlign   = 0              E+00
ned       = 5.35E-03
```

Output from *Anopt*. When *Anopt* runs, the screen output gives information on how the run is progressing. In the screen output from the previous run we see several columns. A brief explanation of their meaning follows.

It	Iteration number
Value	$\sup_e \Gamma(e, f(e))$, the value at current iteration.
Step	sup norm of difference between last two iterates f.
Flat	Flatness optimality diagnostic. It is zero at the solution.
GradAlign	Gradient alignment optimality diagnostic. It is zero at the solution.
ned	A measure of numerical noise in calculations.
Sm.	Indicates whether smoothing takes place.
Grid	Number of points on the unit circle used for function evaluation.

In our run above, Flat went from 0.99 down to 0.018 in three iterations, while GradAlign was zero throughout the iteration (this is not unusual for scalar-valued examples). At the moment of stopping, the iteration was making acceptable progress, but the (large) error tolerance we gave as input prevented the program *Anopt[]* from obtaining more accurate results.

Solution. When the run is over you will have access to the calculated value of the function f^*, under the name **Solution**. This is a list of lists with the following format:

$$\text{Solution} = \{\{x_1, x_2, \ldots, x_n\}, \{y_1, y_2, \ldots, y_n\}, \ldots, \{z_1, z_2, \ldots, z_n\}\},$$

where $x_1, x_2, \ldots, z_n$ are complex numbers. The kth entry of Solution (itself a list) is Solution[[k]].

```
Dimensions[Solution]
{1,  32}
```

Thus in our example, only one scalar-valued analytic function was produced. It consists of 32 values that are the result of sampling the function on a grid of 32 equally spaced points in the unit circle.

Plotting solutions. We now plot the solution as a discrete curve on the complex plane, using the command

```
DiskListPlot[Solution]
```

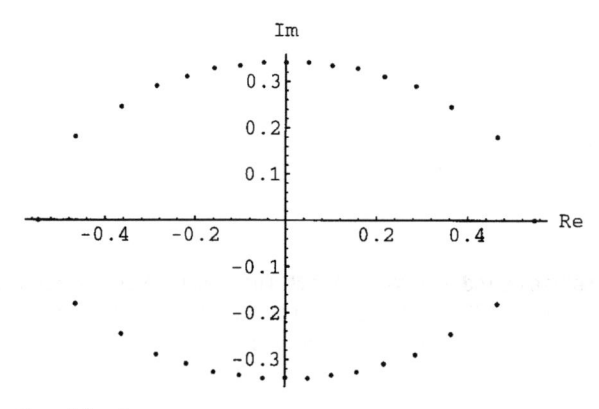

-GraphicsArray-

You can oriduce plots or manipulate the output algebraically. Also, the package *Anopt.m* comes with a function for displaying the solution in 3-D.

```
DiskListPlot3D[Solution]
```

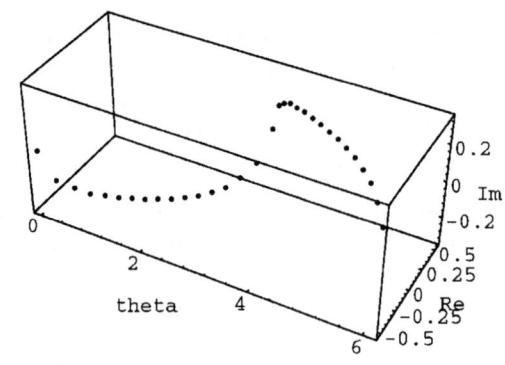

-GraphicsArray-

D.4 Example 2: The case of vector-valued analytic functions

The problem. We consider now a problem of optimization over A_2, which is the space of pairs (f_1, f_2) of analytic functions. We want to find f^* that solves OPT for the performance function

$$\Gamma(e, z_1, z_2) = \text{Re}[1/e + z_1]^2 + 4\,\text{Im}[1/e + z_1]^2 + \text{Re}[1/e + z_2]^2 + 0.3\,\text{Im}[1/e + z_2]^2.$$

Input. To translate this to the Mathematica language, proceed as before and replace z_1 by $z[1]$, z_2 by $z[2]$, etc. The set of inputs is

```
g = Re[1/e + z[1]]^2  +  4   Im[1/e + z [1]]^2
        + Re[1/e + z[2]]^2 + 0.3 Im[1/e + z[2]]^2;

<< Anopt[g,0.001];
```

```
It :   Current Value    : Step      : Optimality Tests  : Error   : Sm. : Grid
   :   gammastar        :           : flat     grAlign  : ned     :     :
-------------------------------------------------------------------------------
0 : 4.3               E+00 : N/A     : 5.3E-01 : 7.2E-01D : N/A    : NON : 32
1 : 2.1545988017648E+00 : 7.8E-01 : 3.2E-06 : 5.5E-02D : 3.2E-05 : NON : 32
2 : 2.1485374717714E+00 : 7.6E-02 : 1.5E-03 : 1.2E-02D : 5. E-05 : NON : 32
3 : 2.1469189941809E+00 : 7.5E-03 : 3.7E-06 : 5.9E-04D : 5.3E-06 : OA5 : 32


Summary
-------
gammastar = 2.146918994180885E+00
flat      = 3.6801434547E-06
grAlign   = 5.9442841631E-04
ned       = 5.3 E-06
```

Stopping, and how good is the solution? The run stops when the equalities

```
Flat < tol,   GrAlign < tol
```

are satisfied, where tol is the error tolerance set by the user. At the solution one must have that Flat = GradAlign = 0, so small Flat and small GradAlign is a necessary condition for a calculated guess at the answer to be close to the actual solution.

Sometimes there is difficulty in the calculations. *Anopt[]* has basically two ways to deal with it: smoothing and grid size doubling. Both are performed automatically if the internal algorithms of *Anopt* indicate that such action is necessary. The user may supress smoothing or grid doubling by specifying options when *Anopt* is run. In the run above, smoothing occurred at the third iteration, while the grid size remained constant at 32.

Fourier coefficients of the solution. Computing the (discrete) Fourier coefficients of functions is useful for many purposes. The Mathematica command Fourier[list], where list = $\{x_1, x_2, \ldots, x_n\}$, is an implementation of the FFT algorithm and returns a list of the Fourier coefficients (times the numerical factor Sqrt[n], where n = number of sample points). To illustrate the use of Fourier[], we compute the Fourier coefficients of Solution[[2]] and store them under the name "fc2":

```
fc2 = Fourier [ Solution [2] ] /Sqrt[32.];
```

The resulting coefficients are ordered according to the indices in $\{0, -1, -2, -3, \ldots, 3, 2, 1\}$. We see below that there is only one nontrivial Fourier coefficient in the first entry of Solution:

```
Chop[ fc2 ]
{0, 0, 0, 0, 0, 0, 0, 0, 0, 0, 0, 0, 0, 0, 0, 0,
-(5.936871299232548*10^-9), 0, 2.481559157373434*10^-8, 0,
5.036991796409501*10^-9, 0 , 1.020976750493356*10^-9,
0, 2.063555217982383*10^-10, 0, 0, 0, 0, 0, -0.6819202501387195}
```

Note that it is clear from the expression above that the function f2 of the calculated solution $f = (f1, f2)$ is a multiple of $e^{i\theta}$.

D.5 Example 3: Specification of more input

Anopt is based upon an iterative scheme that generates a sequence of (discrete) analytic functions f_1, f_2, f_3, etc. This process begins with an initial guess f_0, which in examples 1 and 2 has been defaulted to the function zero, on a discrete grid of 16 points. In this section you will learn how to start *Anopt* on an initial guess f_0 that you specify as a rational function sample on the circle.

Statement of the problem: Example 1 revisited. Suppose we want to solve example 1, only we want to start the iteration with the initial function $f(e) = (e^2 - 3e + 4)/(e + 3)$. We define the performance first.

```
g= Abs[0.8 + (1/e + z[1])^2]^2;
```

We can specify the initial guess at solution as an express in "e" (which represents Exp[I theta]). For example, we set

```
f0 = {  (e^2-3e+4)/(e+3)  } ;
```

The performance function we deal with here has only one complex variable ($N = 1$); hence only one entry is required in f_0. In the case $N > 1$ the different entries of f_0 have to be separated by commas.

The calling sequence to be used now takes as input the function f_0 as initial guess, besides the performance g and the tolerance tol. The sequence is Anopt[g, tol, f_0]. Many other inputs can be specified as options. Below we set the maximum number of iterations to be 3:

D.5. EXAMPLE 3: SPECIFICATION OF MORE INPUT 133

```
Anopt[g, .0001,f0,Iterations->3]
```

It	Current Value gammastar	Step	Optimality Tests flat grAlign	Error ned	Sm.	Grid
0	9.6039969948935E+01	N/A	9.3E-01 : 1. E+00	N/A	NON	32
1	1.9365155086078E+00	3.8E+00	9.1E-01 : 0 E+00	2. E-02	NON	32
2	1.1423579024226E+00	2.8E-01	2.4E-01 : 0 E+00	3.9E-04	5A10	32
3	1.0093525941247E+00	6.7E-02	2.1E-02 : 0 E+00	2.1E-03	NON	32

```
Summary
-------
gammastar = 1.009352594124695E+00
flat      = 2.08995166745E-02
grAlign   = 0            E+00
ned       = 2.12E-03
```

Another valid way to specify the initial guess is a list of values obtained by sampling the function on a grid of equally spaced points. To produce a discrete version of f_0 on a 32-point grid you can use a replacement rule. Note the braces around the rational function:

```
f0 = { (e^2-3e+4)/(e+3) } /.
          {e->Table[N[Exp[2 Pi I i/32]],{i,0,31}]};
```

Restarting the run. In the above, the process stopped because the limit on the number of iterations was attained. If you want *Anopt* to proceed from the last iteration above, you can restart the iteration. For this take as an initial guess for the new run the last iterate of the previous run, which is stored in the variable Solution.

Below we double the grid size of the answer we obtained in the run above and assign it to $f1$. Then $f1$ is used to restart the iteration.

```
f1 = DoubleGrid[Solution];
Anopt[g,0.00002,f1]
```

It	Current Value gammastar	Step	Optimality Tests flat grAlign	Error ned	Sm.	Grid
0	1.0093525941247E+00	N/A	2.1E-02 : 0 E+00	N/A	NON	64
1	1.0005560979092E+00	5.7E-03	1.1E-03 : 0 E+00	2.3E-03	NON	64
2	1.0002780415569E+00	9. E-04	5.6E-04 : 0 E+00	1.2E-03	5A10	64
3	1.000033365362 E+00	1.6E-04	6.7E-05 : 0 E+00	2. E-03	NON	64
4	1.0000050181354E+00	1.6E-05	1.3E-05 : 0 E+00	3.1E-03	NON	64

```
Summary
-------
gammastar = 1.000005018135407E+00
flat      = 1.291679907 E-05
grAlign   = 0            E+00
ned       = 3.1 E-03
```

The solution has the following structure.

```
Dimensions[Solution]
{1, 64}
```

Manipulating solutions. As an example of how to manipulate output, we plot below the discrete values of the function $\Gamma(e, f^*(e))$. We then compare this graph with the claim (from the theory) that gdisc(e) is constant. For this one must substitute in the formula $g = g(e, z[1])$ the variable $z[1]$ by Solution[[1]] and e by its discrete values on a grid of equally spaced points. This is accomplished using replacement rules as shown below. First we produce a discrete version of Exp[I theta] for theta in $[0, 2\,Pi]$.

```
ngrid = 64
edisc = Table[N[Exp[2  Pi  I  (i-1)/ngrid]],{i,1,ngrid}];
```

Now we use replacement rules.

```
gdisc = g /. {e -> edisc, z[1]->Solution[[1]]};
```

The plot is produced below. Here we specify a range:

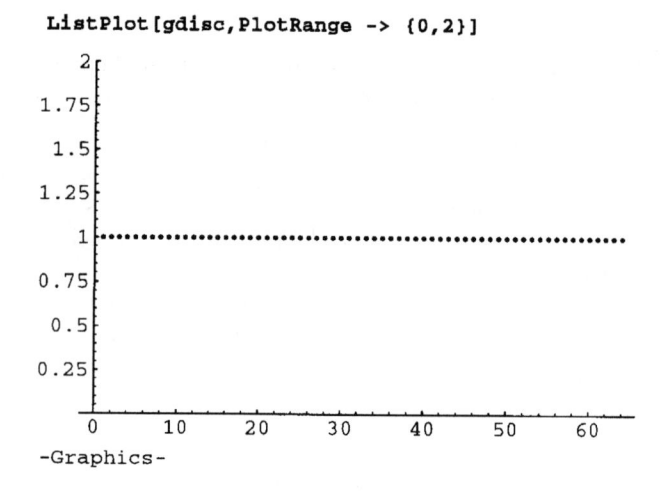

ListPlot[gdisc,PlotRange -> {0,2}]

-Graphics-

We conclude that $g1$ is approximately constant. Let us find out how close to constant:

```
    Max[gdisc]-Min[gdisc]
0.0000129169
```

Combining algebra and plots. Now we try something a bit different. Suppose you wish to compute and plot in 3-D the partial derivative of the performance with respect to z, where z = our calculated solution. This requires only two lines of input for the user. We also plot the system on the complex plane for comparison. The function grad defined below is discrete; that is, it is a list of values.

```
grad   = ComplexD[g,z[1]] /.
          {e -> edisc,  z[1]->Solution[[1]]};
```
DiskListPlot[grad,PlotJoined->True]

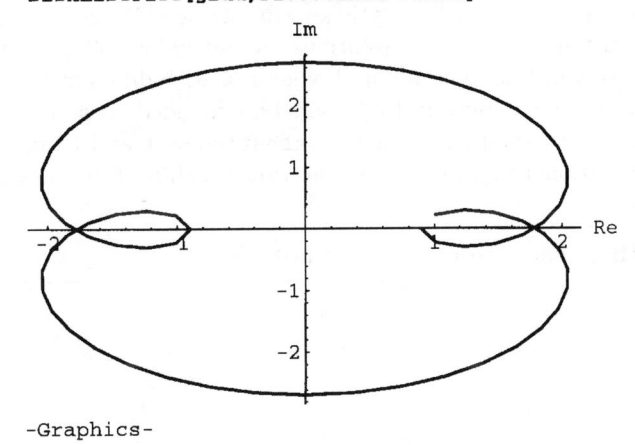

[Plot showing ellipse-like curve in complex plane with Re and Im axes, ranging approximately -2 to 2]

-Graphics-

D.6 Quick reference for *Anopt*

The output table. After every iteration, eight numbers are placed on the screen. Recall that to each new function f_k generated by *Anopt* correponds to a new line in the output table. The first line (iter= 0) corresponds to f_0. Since we did not specify f_0 in the two examples above, it has been defaulted there to the zero function. The columns of the table are explained below.

HEADING	EXPLANATION
Iter	Iterate number k.
SupGamma	$\sup G(e, f(e))$, the value at current f.
Step	sup norm [difference between last two iterates f].
Flat, GradAlign	Optimality tests. If f is near optimal, then Flat and GradAlign are close to 0.
NED	Numerical Error Diagnostic. If relatively large it indicates numerical trouble, usually due to slowly decaying Fourier coefficients.
Sm	Automatic smoothing of f_k.
Grid	Number of samples on the unit circle, a power of 2. increased by *Anopt* as needed.

Usually the most important numbers on the screen are the Flat and GradAlign diagnostics. As they approach zero, the current guess is expected to approach the solution. Under mild hypotheses, the solution is unique in *OPT* problems with only one (scalar-valued) analytic known function. When the number of analytic unkown functions $N > 1$, solutions are not unique. There may be many local solutions, which *Anopt* may find when run with different initial guesses.

Explanation of the output table will be expanded in section 1. A more advanced user may set the options `Diagnostics-> 1` and `Diagnostics2->1`. This gives additional information on the run, which is stored automatically in certain files.

Some Mathematica functions and notation

Abs[z]	Absolute value of z.
Conjugate[z]	Conjugate of z.
z^n	nth power of z.
Re[z]	Real part of z.
Im[z]	Imaginary part of z.
{x1,x2,x3}	A list with entries $x1, x2, x3$.
D[g,z]	Derivative of g with respect to z.
Fourier[f]	Fast Fourier transform of f.

Some *Anopt* utilities

ComplexCoordinates[g]	Applies transformation rules to g to write in terms of the complex variables and their conjugates.
ComplexD[g,z]	Complex derivative of g with respect to z.
ComplexD[g,Conjugate[z]]	Complex derivative of g with respect to Conjugate $[z]$.
DoubleGrid[f]	Doubles the number of sample points of a discrete function.
HalfGrid[f]	Halves the number of sample points of a discrete function.
DiskListPlot[f]	Plots a list of complex numbers in the Re-Im plane.
DiskListPlot3D[f]	Plots a list of complex numbers in 3-D space. One of the axes corresponds to the parameter, assumed to be theta in [0, 2 Pi].

Appendix E

NewtonInterpolant Notebook

Interpolation with rational functions using NewtonInterpolant

J.W. HELTON AND O. MERINO

Lab. for Mathematics and Statistic, University of California, San Diego

Introduction

Let $R[s]$ be a function of the complex variable s, and let

$$\{s_1, s_2, \ldots, s_n\}$$
$$\{z_1, z_2, \ldots, z_n\}$$

be given sets of n complex numbers. The relation $R[s1] = z1$ is called the interpolation condition, and several of these form a "set of interpolation conditions":

$$R[s1] = z1, \quad R[s2] = z2, \ldots, R[sn] = zn \qquad (INT)$$

A function $R[]$ that satisfies *(INT)* is called an *interpolant* for *(INT)*.

The function NewtonInterpolant[] produces a rational function, say $R[s]$, that satisfies a set of interpolation conditions specified by the user as input, and has all its poles at a single location, typically a negative real number.

There are many rational functions that satisfy *(INT)*. NewtonInterpolant[] produces one that is the most economical in the sense that it is proper and the denominator has the smallest degree possible.

NewtonInterpolant[] can also deal with more general problems, with interpolation conditions on the derivatives of the function.

E.1 First calculation of an interpolant

Begin by loading the package

In[6]:=

```
<<NewtonInterpolant`
```

The problem is to find an interpolant $R[s]$ for the interpolation conditions

$$(INT) \qquad R[3] = 1, \quad R[1] = -2$$

The answer is obtained with

In[8]:=

```
data = { {3,1} , {1,-2} };
rat1 = {NewtonInterpolant[data,s]
```

Out[8]=

```
  4 (-2 + s)
  ------------
     1 + s
```

To verify that the rational function is in fact an interpolant is easy:

In[9]:=

```
rat1 /. s->3
```

Out[9]=

```
  1
```

E.2 Specifying the pole location

In the example above the pole location of the interpolant is $s0 = -1$. You can change this value by giving a pole location as input. For example, suppose you want pole location -4 for the interpolant. This is what you type:

In[10]:=

```
rat2 = NewtonInterpolant[data,s,PoleLocation->-4]
```

Out[10]=

```
  -37 + 17 s
  -----------
   2 (4 + s)
```

E.3 Specifying of the relative degree

If $R = N/D$ is a rational function with N as numerator and D as denominator, the relative degree of R is the integer

$$d(R) = \text{degree}(N) - \text{degree}(D).$$

Functions with nonnegative relative degree are called proper, and if the relative degree is positive the function is called strictly proper.

The user can specify a relative degree as input for NewtonInterpolant[]. Suppose that you want to determine an interpolant for

$$(INT) \qquad R[3] = 1, \quad R[1] = -2$$

with pole location -4 and relative degree 2. This is how you can do it:

In[12]:=

```
rat2 = NewtonInterpolant[data,s,PoleLocation->-4,RelativeDegree->2]
```

Out[12]=

```
-1093 + 593 s
-------------
            3
   2 (4 + s)
```

E.4 Complex numbers as data

Rational functions that appear in engineering have real coefficients. This corresponds in terms of zeros and poles to the statement that complex zeros and complex poles come in conjugate pairs. These functions are called real-rational, and they have the property

$$(RR) \qquad R[\text{Conjugate}[s]] = \text{Conjugate}[R[s]].$$

If some of the data you use to set up an interpolation problem is complex and you want an answer that is real-rational, then you must include pairs of data points that reflect property *(RR)*. For example, if one interpolation condition is

$$R[3 + 2I] = 1 - 5I,$$

in order to obtain a real-rational interpolant, your set of interpolation conditions must include both

$$R[3 + 2I] = 1 - 5I \quad \text{and} \quad R[3 - 2I] = 1 + 5I.$$

Note: If one of the conditions is not given as input, NewtonInterpolant[] will give as interpolant a non–real-rational, since it will assume that you are willing to give up the condition of real-rational in order to produce the most "economical" answer.

Problem. Find a real-rational function R such that $R[I] = 0, R[2+3I] = 4-2I, R[1] = 3$.

Solution. First expand the set of interpolation conditions to

(INT) $\quad R[I] = 0, \quad R[-I] = 0, \quad R[2+3I] = 4-2I, \quad R[2-3I] = 4+2I, \quad R[1] = 3$

and follow with

In[15]:=

```
data = {  {I,0},{-I,0},{2 + 3 I,4 - 2 I},
          {2 - 3 I,4 + 2 I},{1,3}  };

rat4 = NewtonInterpolant[data,s]
```

Out[15]=

```
                2       3    4
3 (-4 + 13 s - 5 s + 13 s - s )
--------------------------------
                4
          (1 + s)
```

E.5 Higher-order interpolation

More general interpolation conditions are produced when, besides the value of the function at a point, values of the derivatives of this function at the point are specified. NewtonInterpolant can handle this case too.

Problem. Find an interpolant R such that $R[1] = -3$, $R[2] = 0$, and $R'[2] = 1$.

In[17]:=

```
data = {  {1,-3} ,  {2,0,1}  };
rat = NewtonInterpolant[ data, s]
```

Out[17]=

```
              2
3  (-10 + 7 s - s )
-------------------
          2
    (1 + s)
```

E.5. HIGHER-ORDER INTERPOLATION

Problem. Find a real-rational interpolant R with relative degree 1 such that
$$R[2I] = 0 \quad \text{and} \quad R'[2I] = 1 - 3I.$$

Solution. In this case it is necessary to consider pairs of complex data as follows.

In[19]:=

```
data = {  {2 I,0,1 - 3 I},{-2 I,0,1 + 3 I}  };
NewtonInterpolant[ data ,s, RelativeDegree -> 1]
```

Out[19]=

```
                   2       3
-24 + 316 s  -  6 s  + 79 s
-----------------------
                 4
        8 (1 + s)
```

Appendix F

NewtonFit Notebook

NewtonFit

Introduction

This Mathematica notebook is the documentation for the package NewtonFit, for treating nonlinear L2 data fitting problems. The main function is NewtonFit[], which is an implementation of the Newton algorithm.

Acknowledgment. Discussions with Jim Easton were helpful.

An L2 approximation problem

The L2 optimization or approximation problem considered here is

(Prob)   Given a function $f(x, a1, a2, \ldots, am)$ and two sets of n numbers each,
(data points) $P = \{x1, x2, x3, \ldots, xn\}$
(data values) $V = \{y1, y2, y3, \ldots, yn\}$,
Find (if it exists) an m-tuple $(a1, a2, a3, \ldots, am)$ that minimizes

$$\chi 2(a1, a2, \ldots, am) = \sum_{k=1,n} (yk - F(xk, a1, a2, \ldots, am))^2$$

The Mathematica function Fit[] can be used to treat *(Prob)* when the function $F(x, a1, \ldots)$ is linear in the a's. For the general problem *(Prob)* a more powerful algorithm is necessary. One example of this is the classical Newton method, implemented here as the function NewtonFit. Another is the well-known Gauss-Newton algorithm. These are iterative procedures that, when they converge, produce a local solution to *(Prob)*. This is the best one can hope for for such a general optimization problem. In practice, one way to go about finding global solutions is to run the algorithm repeatedly with different initial guesses, with the hope that the global solution will be found by one of these runs.

The emphasis in these notes is on models $F(x, a1, \ldots, am)$ that are rational functions in x, with coefficients given by $a1, a2, a3, \ldots, am$. The algorithm implemented here is very general in that, in principle, it solves *(Prob)* for F rational and for many other functions not necessarily rational. The wide range of problems that can be treated with this implementation is due to the fact that Mathematica can do both symbolic and numerical calculations.

What does NewtonFit[] do for you? The package NewtonFit.m contains the function `NewtonFit[]`. You can

a. Find local solutions to standard nonlinear L2 data fitting problems for a wide range of models, in particular, models that are rational functions of given order.

b. Solve weighted nonlinear L2 data fitting problems.

c. Find rational approximations to the data, where the degree and the *minimum number of stable zeros and poles* is specific a priori.

Problems you may encounter when using NewtonFit when treating rational models

1. `NewtonFit` will fail frequently in arriving at a local solution. The nature of the problem of approximation with rational models is such that for many cases there are many narrow and curved valleys in parameter space. This produces a high degree of instability in local algorithms, such a Newton and Gauss-Newton. A good initial guess is very important.

2. Even if a local solution is produced you can never be sure if it is a global solution (unless the model is linear in the parameters, i.e., the classical linear fit).

In practice, a partial cure for both problems 1 and 2 is to produce lots of runs, initializing each time at a different location in parameter space. More satisfactory would be to complement the local algorithms with different algorithms having better global behavior. This is not done here though, for lack of space.

F.1 First example

In this example, we sample a high-order rational function of "s" on the imaginary axis "Iw" and fit a low-order rational to the data that we obtained.

Consider the function

(1)   `F[s] =1 / (s^3 + 6 s^2 + 11 s +6)`.

F.1. FIRST EXAMPLE

A grid of points $j*w\,i$ on the imaginary asix is given by

(2)   `w[[i+1]]=.001  (1.1)^i,`   where   i=0,..., 99

and corresponding values $v[[i]]$ are generated as

(3)   `v[[i]] = F[ I* w [[i]] ].`

Following Luus and Shenton–Shifiei, a "reduced" second-order model

$$(4) \quad g[s,a] = \frac{a[1]s + 1}{a[2]+a[3]s+a[4]\ s^2}$$

is sought.

Problem. Given datapoints (2), datavalues (3), and the model (4), find parameters a[1],a[2],a[3],a[4] that minimize

(5)   `Sum[  Abs[ v[[i]] - g[ I*w[[i]] ,a] ]^2,   {i,1,n} ]`

Load the package

```
SetDirectory["~/BOOK/CODEUPDATE"];
<<NewtonFit.m;
```

First we produce the gridpoints and grid values:

```
w = Table[0.01*(1.1)^i,{i,0,99}];
F[s_]:=1/(s^3 + 6. s^2  + 11. s + 6.);
values = F[I w];
points = I w;
```

The gridpoints s and gridvalues v are combined as a list of pairs (s,v) called data below. This is one input for `NewtonFit[]`.

```
data = Transpose[{points,values}];
```

The model is set below in terms of parameters $a[1], a[2], \ldots$ and the variable s.

```
model = (a[1]  s + 1.)/(a[2] s^2 + a[3] s + a[4]);
```

We now choose an initial set of parameters to start the iteration. If these are not specified, the function `NewtonFit[]` defaults all of the to 0.

```
initial1 = {-1.,5.,9.,4.};
```

The output of Newton is assigned to the variable output1 below. This will be helpful when manipulating the results of the call to `NewtonFit[]`.

```
output1 = NetwonFit[data,model,s,a,Parameters->initial1]
```

```
-----------------------------------------------
Iter     Step          grad         cost
-----------------------------------------------
0                     0.754419      0.581507
1       4.05838       0.482129      0.313686
2       3.0786        0.0714609     0.0123889
3       2.09594       0.00824722    0.000601043
4       0.389809      0.000208885   0.000327769
                              -6
5       0.0285536     1.66219 10    0.000327117
                                        -11
6       0.0000899072  3.82929 10    0.000327117
```

```
{Parameters -> {-0.134521, 4.86549, 10.0684, 6.01588},
   SingValHessian -> {0.401005, 0.0583707, 0.00385517, 0.00126542},
                                              -11
   Cost -> 0.000327177, NormGradient -> 3.82929 10
   NIterations -> 6}
```

For an explanation of the diagnostics—for example, Sing ValHessian—type ?SingValHessian at the Mathematica promt to get a description.

The results of the run above say that a local minimum has been found, since the gradient of the objective function is close enough to 0, and the hessian has the correct signature. This is how you form a function with the optimal parameters out of the model:

```
par = Parameters /. output1;
r[s_] = model /.{a[j_] :> par[[j]]}
```

```
          1. -0.134521 s
        -------------------------------
                                  2
        6.01588 + 10.0684 s + 4.86549 s
```

A plot is now produced that contains the (discrete) data being approximated and the optimal function r[s].

```
Show[
      ListPlot[Transpose[{Re[values],Im[values]}],
         DisplayFunction->Identity],
      ParametricPlot[{Re[r[I t]],Im[r[I t]]},{t,0,10.},
         DisplayFunction->Identity],
   DisplayFunction->$DisplayFunction];
```

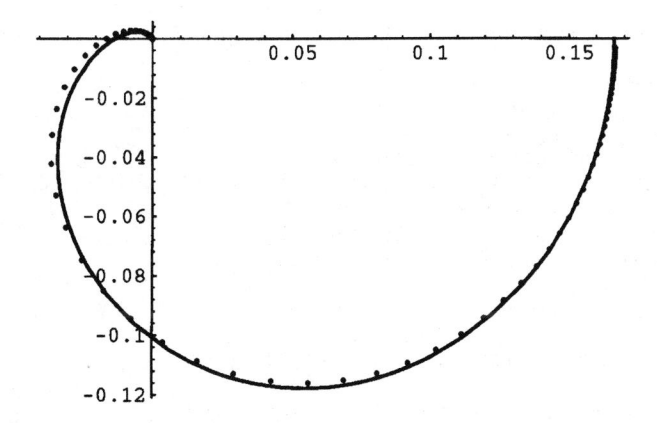

F.2 Template for many runs

The example above is not typical in that the run produced satisfactory results in the first trial. Approximation by rational functions is tricky business; usually there are lots of local minima, and problems due to indefinite hessians are not uncommon. Below you will find an example of a set of commands for running the program many times starting the iteration at different random locations in the space of parameters.

The example treated is the same as above. Initial sets of parameters are generated randomly with values a[i] in the interval $(-5, 5)$. The output of individual runs is stored for later inspection in a file named output1.

```
<<NewtonFit.m;
w = table[0.01*(1.1)^i,{i,0,99}];
F[s_]=1/(s^3 + 6. s^2 + 11. s + 6.);
values = F[I w];
points = I w;
data = Transpose[{points,values}];
model = (a[1] s + 1.)/(a[2] s^2 + a[3] s + a[4]);
Do[
   initial = 10.{Random[],Random[],Random[],Random[]} -5.;
   output = NewtonFit[data,model,s,a,
                          Parameters->initial];
   Save["output1",initial,output];
    ,{i,1,50}];
Exit
```

F.3 Using a weight

It is possible to emphasize some data points over others by means of a weight function as in

```
Sum[ weight[[i]]*( Abs[ v[[i]] - g[ I*w[[i]] ,a] ]^2) , {i,1,n} ]
```

A weight can be specified as input for the function `NewtonFit[]`.

Suppose that the example considered in section 1 gives an answer that we reject as producing too large a deviation from the model to the data for values of w[[i]] i= 25, ..., 45. To reduce the error at these 21 points a weight function is defined as

```
wt = Join[ Table[1.,{24}],Table[10.,{21}],Table[1.,{55}]];
```

We can use the parameter values found in the previous run to initialize the current one.

```
initial2 = Parameters /. output1;
output2 = NetwonFit[data,model,s,a,
         Parameters->initial2,Weight->wt]
```

```
---------------------------------------------
Iter    Step         grad           cost
---------------------------------------------
 0                0.00372873      0.000526309
 1    0.165522    0.0000226568    0.000401831
                              -8
 2    0.00417434  2.57386 10      0.000401817
                -7              -10
 3    3.43866 10  3.06373 10      0.000401817
                              -10
 4    0.          3.06373 10      0.000401817
```

```
{Parameters -> {-0.135428, 4.71745, 10.136, 6.01399},
    SingValHessian -> {1.39787, 0.23132 , 0.00153349, 0.00155875},
                                                              -10
    Cost -> 0.000401817, NormGradient -> 3.06373 10
    NIterations -> 4}
```

To plot the resulting function set, just proceed as in the first example.

F.4 Stable zeros and poles

A zero z of a rational function $R[x]$ is stable if $\text{Re}[z] < 0$. A pole z of $R[x]$ is stable if $\text{Re}[z] < 0$. We now consider the problem of finding the set of optimal parameters in *(Prob)* with rational model, with the additional constraint that the resulting rational function has either (a) stable poles, (b) stable zeros, and (c) stable poles and zeros. The key in solving this problem with the tools we have is the following result.

PROPOSITION. *Let $P[x]$ be a monic polynomial with real coefficients. Then $P[x]$ has all its zeros in the left open (closed) left half-plane if an only if when P is factored over the reals as a product of degree 1 and degree 2 monic polynomials, the coefficients of each factor are positive (nonnegative).*

F.4. STABLE ZEROS AND POLES 149

We can take advantage of this result because the function `NewtonFit[]` calculates derivatives by symbolic differentiation. As an example, consider the problem from section 1 with the additional constraint that all zeros and poles of the resulting rational function are located in the left half-plane. Note that the answer obtained in section 1 is not satisfactory, because it has a zero in the RHP. Define as model

```
model3 = a[4]*(a[1]^2 s + 1.)/(a[2]^2 s^2 + a[3]^2 s + 1);
```

Note that any choice of parameters in model3 produces rational functions with zeros and poles off the right-hand plane.

The set of initial parameters given below has been determined with a multiple run of `NewtonFit[]` with several initial sets of parameters. This set in particular produces a local solution as shown below.

```
initial3 = {0.,.9,1.3,.17};
output3 = NetwonFit[data,model3,s,a,
                   Parameters->initial3]
```

```
-----------------------------------------------
Iter    Step            grad            cost
-----------------------------------------------
0                       0.323703        0.00430163
1       0.112943        0.022005        0.00156777
2       0.00762118      0.0000177771    0.00154469
                                    -9
3       0.0000305235    1.18668 10      0.00154469
                -11                 -9
4       4.53645 10      1.08998 10      0.00154469

{Parameters -> {0., 1.00936, 1.31714, 0.166463},
    SingValHessian -> {91.4903, 0.77512, 0.350922, 0.0313337},
                                                -9
    Cost -> 0.00154469, NormGradient -> 1.08998 10
    NIterations -> 4}
```

Now the rational function obtained above is produced, and a plot is generated.

```
par3 = Parameters /. output3;
r3[s_]= model3 /.{a[j_] :>par3[[j]]}
```

```
0.166463 (1. +0. s)
-------------------
                        2
1 + 1.73484 s + 1.01881 s

Show[
        ListPlot[Transpose[{Re[values],Im[values]}],
            DisplayFunction->Identity],
```

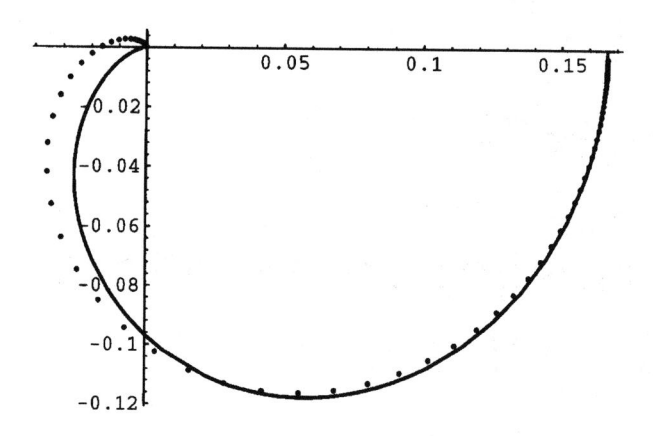

```
    ParametricPlot[{Re[r3[I t]],Im[r3[I t]]},{t,0,10.},
        DisplayFunction->Identity],
  DisplayFunction->$DisplayFunction];
```

References

Z. Shafiei and A.T. Shenton, *Theory and Application of H-infinity Disk Method*, Report no. MES/ATS/BAE/002/90, Department of Mechanical Engineering, University of Liverpool, U.K.

R. Luus, *Optimization in model reduction*, Int. J. Control, 32 (1980), pp. 741–747.

P. Gill, Q. Murray, and M. Wright, *Practical optimization*, Academic Press New York, 1986.

Appendix G

OPTDesign Plots, Data, and Functions

Some users may want to manipulate the output of an OPTDesign run before dealing with rational fits. We present examples of OPTDesign commands to plot and manipulate T, L, Co, or other lists of data.

Run the example from Chapter 5 before proceeding with the rest of the notebook.

G.1 Functions and grids

Output functions of OPTDesign runs

A run of OPTDesign produces the calculated closed loop T, the open loop L, and the compensator Co as lists. If you type, say, T in a session after you run OPTDesign, then Mathematica returns a list of complex numbers that are the values of the calculated closed-loop function values on a w-axis grid.

Note that the grid points on the w axis do not appear explicitly when you ask Mathematica to show T.

```
In[17] := T
Out[17] = {0, -0.0371551 + 0.0341142 I, -0.124852 + 0.00349232 I, -0.184928 - 0.129109 I,
      -0.114989 - 0.306715 I, 0.0780776 - 0.371381 I, 0.212942 - 0.285464 I,
      0.236439 - 0.199328 I, 0.231823 - 0.160042 I, 0.230235 - 0.141843 I, 0.230142 - 0.133257 I,
      0.233164 - 0.136253 I, 0.250685 - 0.152637 I, 0.301352 - 0.162629 I, 0.365523 - 0.131061 I,
      0.40342 - 0.0669143 I, 0.41222 + 0.   I, 0.40342 + 0.0669143 I, 0.365523 + 0.131061 I,
      0.301352 + 0.162629 I, 0.250685 + 0.152637 I, 0.233164 + 0.136253 I, 0.230142 + 0.133257 I,
      0.230235 + 0.141843 I, 0.231823 + 0.160042 I, 0.236439 + 0.199328 I, 0.212942 + 0.285464 I,
      0.0780776 + 0.371381 I, -0.114989 + 0.306715 I, -0.184928 + 0.129109 I,
      -0.124852 - 0.00349232 I, -0.0371551 - 0.0341142 I}
```

APPENDIX G. OPTDESIGN PLOTS, DATA, AND FUNCTIONS

What is the grid you are currently using in an OPTDesign session?

The assumption is that there is a grid that is used for sampling all functions of the OPTDesign session. To see it, type

```
In[18] := Grid[]
```
Out[18] = {∞, 10.1532, 5.02734, 3.29656, 2.41421, 1.87087, 1.49661, 1.2185, 1., 0.820679, 0.668179, 0.534511, 0.414214, 0.303347, 0.198912, 0.0984914, 0., -0.0984914, -0.198912, -0.303347, -0.414214, -0.534511, -0.668179, -0.820679, -1., -1.2185, -1.49661, -1.87087, -2.41421, -3.29656, -5.02734, -10.1532}

Putting together the grid and the values of a function

If you wish, you may produce a list of pairs of the form $\{w, T[Iw]\}$. To do this, type

```
In[19] := Tpairs = OPTDParametrize[T]
```
Out[19] = {{∞, 0}, {10.1532, -0.0371551 + 0.0341142 I}, {5.02734, -0.124852 + 0.00349232 I},
{3.29656, -0.184928 - 0.129109 I}, {2.41421, -0.114989 - 0.306715 I},
{1.87087, 0.0780776 - 0.371381 I}, {1.49661, 0.212942 - 0.285464 I},
{1.2185, 0.236439 - 0.199328 I}, {1., 0.231823 - 0.160042 I},
{0.820679, 0.230235 - 0.141843 I}, {0.668179, 0.230142 - 0.133257 I},
{0.534511, 0.233164 - 0.136253 I}, {0.414214, 0.250685 - 0.152637 I},
{0.303347, 0.301352 - 0.162629 I}, {0.198912, 0.365523 - 0.131061 I},
{0.0984914, 0.40342 - 0.0669143 I}, {0., 0.41222 + 0. I},
{-0.0984914, 0.40342 + 0.0669143 I}, {-0.198912, 0.365523 + 0.131061 I},
{-0.303347, 0.301352 + 0.162629 I}, {-0.414214, 0.250685 + 0.152637 I},
{-0.534511, 0.233164 + 0.136253 I}, {-0.668179, 0.230142 + 0.133257 I},
{-0.820679, 0.230235 + 0.141843 I}, {-1., 0.231823 + 0.160042 I},
{-1.2185, 0.236439 + 0.199328 I}, {-1.49661, 0.212942 + 0.285464 I},
{-1.87087, 0.0780776 + 0.371381 I}, {-2.41421, -0.114989 + 0.306715 I},
{-3.29656, -0.184928 + 0.129109 I}, {-5.02734, -0.124852 - 0.00349232 I},
{-10.1532, -0.0371551 - 0.0341142 I}}

Let's check the size of the list *Tpairs*.

```
In[20] := Dimensions[Tpairs]
```
Out[20] = {32, 2}

G.2 Plots

Plotting the envelope and T simultaneously

```
In[21] := EnvelopePlot3D[Radius -> r0, Center -> k0, ClosedLoop -> T];
```

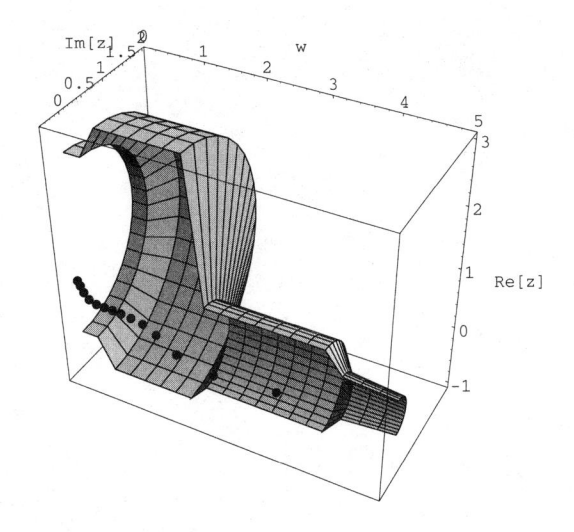

The points of T may be joined by typing

```
In[22] := EnvelopePlot3D[Radius -> r0, Center -> k0, ClosedLoop -> T, PlotJoined -> True];
```

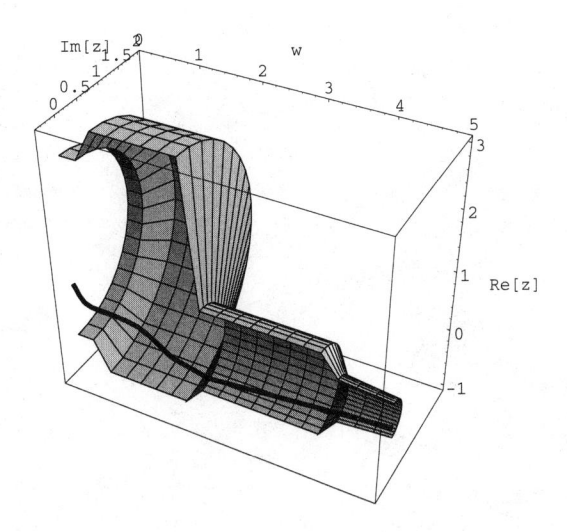

Bode plots of the discrete function T

The commands BodeMagnitude[T] and BodePhase[T] take as input either functions defined by formulas or lists of data.

```
In[23] := BodeMagnitude[T,FrequencyBand -> {0.01,10.}];
```

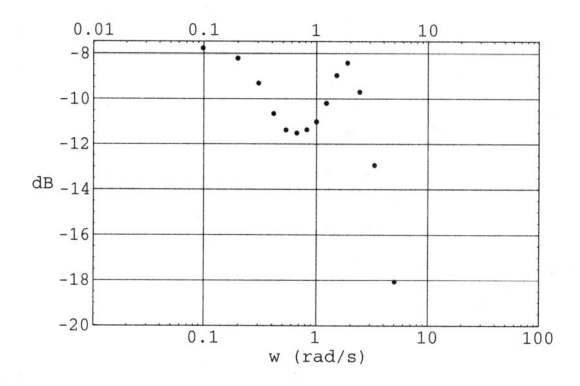

```
In[24] := BodePhase[T,FrequencyBand -> {0.1,10.}];
```

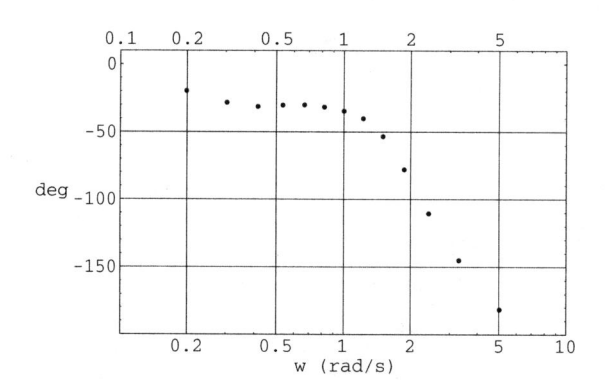

3-D list plot of gridvalues of T

To plot in 3-D discrete functions of frequence (i.e., lists of values such as T), use the command

G.2. PLOTS

In[25] := RHPListPlot3D[T,PlotRange -> {{0,3},Automatic,Automatic}];

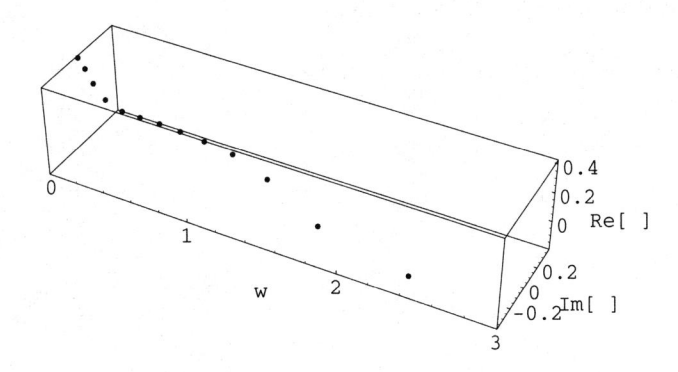

Nyquist plot definition and examples

The Nyquist plot is produced with

In[26] := Nyquist[L];

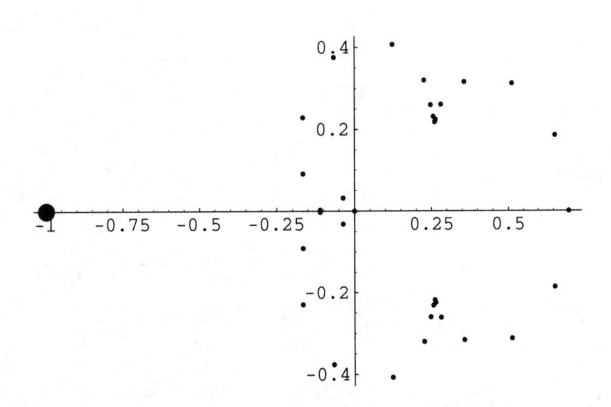

The points can be joined by typing the command

```
In[27] := Nyquist[L, PlotJoined -> True];
```

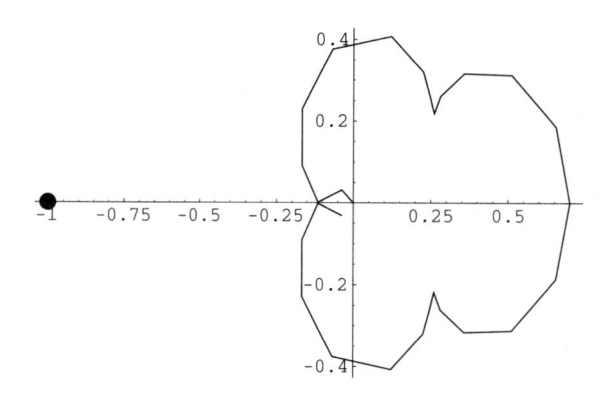

Nichols plot of gridvalues

Here is a Nichols plot of the discrete function L.

```
In[28] := Nichols[L, PlotJoined -> True];
```

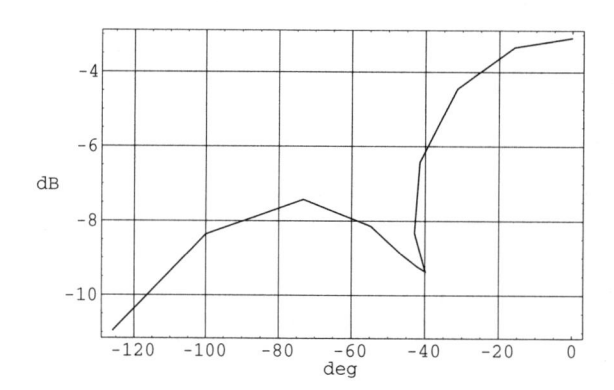

G.3 Rational approximation and model reduction

From data on the grid to rational functions

To find a stable rational function that corresponds to a closed loop function T generated by OPTDesign (and that satisfies the internal stability requirements), you may use

G.3. RATIONAL APPROXIMATION AND MODEL REDUCTION

```
In[29] := Trat = RationalModel[s,DegreeOfDenominator -> 3]

        Error = 0.115867
```

$$\text{Out[29]} = \frac{-5.+s}{(2.+s)^2} + \frac{(-5.+s)(1.50243+1.21125s)}{(-1.21489-0.785108s)(2.+s)^2}$$

A general-purpose Caratheodory–Fejer approximation by stable rationals is implemented as the function StableFit. In addition to the function values, you must specify a grid with a standard OPTDesign format as input (which you can generate outside of OPTDesign if you like) and the degree of the denominator. Here we use the list $T1$ that is a by-product of an OPTDesign run.

```
In[30] := Dimensions[T1]

Out[30] = {32}

In[31] := wpts = Grid[];

In[32] := StableFit[T1,wpts,DegreeOfDenominator -> 3]

        Error = 0.00965834
```

$$\text{Out[32]} = \frac{1059463+4.91307s+3.11098s^2+1.21406s^3}{-1.19052-3.91914s-2.06137s^2-0.828967s^3}$$

For more on rational approximation with a different algorithm, see the NewtonFit notebook (Appendix F).

From a rational function to data on a grid

If you have a rational function Tr rather than a list of values, the following command gives a list of values of Tr on the OPTDesign session grid.

```
In[33] := Tdisc = Discretize[Trat]

Out[33] = {-5.35865 × 10⁻¹⁸, -0.0456089 + 0.0339105 I, -0.125632 - 0.0169765 I, -0.148543 - 0.13693 I,
          -0.0892608 - 0.252746 I, 0.0169445 - 0.316023 I, 0.126378 - 0.322324 I,
          0.213756 - 0.289865 I, 0.27219 - 0.239305 I, 0.304969 - 0.185478 I, 0.318879 - 0.136635 I,
          0.320734 - 0.0962158 I, 0.316079 - 0.0647563 I, 0.30898 - 0.0412528 I,
          0.302207 - 0.0239817 I, 0.29751 - 0.0109313 I, 0.295844 + 0. I, 0.29751 + 0.0109313 I,
          0.302207 + 0.0239817 I, 0.30898 + 0.0412528 I, 0.316079 + 0.0647563 I,
          0.320734 + 0.0962158 I, 0.318879 + 0.136635 I, 0.304969 + 0.185478 I, 0.27219 + 0.239305 I,
          0.213756 + 0.289865 I, 0.126378 + 0.322324 I, 0.0169445 + 0.316023 I,
          -0.0892608 + 0.252746 I, -0.148543 + 0.13693 I, -0.125632 + 0.0169765 I,
          -0.0456089 - 0.0339105 I}
```

References

[AAK68]   V. M. ADAMJAM, D. Z. AROV, AND M. G. KREIN, *Infinite Hankel matrices and generalized problems of Caratheodory–Fejer and F. Riesz* Functional Anal. Appl., 2, (1968), pp. 1–18.

[AAK72]   V. M. ADAMJAM, D. Z. AROV, AND M. G. KREIN, *Analytic properties of Schmidt pairs for Hankel operator and the generalized Schur–Takagi problem*, Math. USSR-Sb., 15 (1972), pp. 15–78.

[AAK78]   V. M. ADAMJAM, D. Z. AROV, AND M. G. KREIN, *Infinite block Hankel matrices and related extension problems*, Amer. Math. Soc. Trans., 111 (1978), pp. 133–156.

[Ag$_{report}$]   J. AGLER, *Interpolation*, preprint, late 1980s.

[Ahl66]   L. V. AHLFORS, *Complex Analysis,* 2nd ed., McGraw-Hill, New York, 1966.

[AHO96]   F. ALIZADEH, J.-P. A. HAEBERLY, AND M. L. OVERTON, *Primal-Dual Interior Point Methods for Semidefinite Programming: Convergence Rates, Stability, and Numerical Results*, Technical report, Computer Science Department, Courant Institute of Mathematical Sciences, New York University, New York, 1996.

[Al96]   H. ALEXANDER, *H. Gromov's method and Benniquin's problem*, Invent. Math., 125 (1996), no. 1, pp. 135–148.

[A63]   T. ANDO, *On a pair of commuting contractions*, Acta Sci. Math. (Szeged), 24 (1963), pp. 88–90.

[BHMer94]   F. N. BAILEY, J. W. HELTON, AND O. MERINO, *Alternative approaches in frequency domain design of single loop feedback systems with plant uncertainty*, Int. J. Robust and Nonlinear Control, submitted.

[**BGR90**] J. A. BALL, I. GOHBERG, AND L. RODMAN, *Interpolation of rational matrix functions*, Birkhäuser-Verlag, Basel, 1990.

[**B92**] H. BEKE, *Algorithm GKL; Documentation for Computer Code GKL*, Department of Electrical Engineering, University of Minnesota, 1992.

[**BHM86**] J. BENCE, J. W. HELTON, AND D. E. MARSHALL, *Optimization over H-infinity*, Proc. Conference on Decision and Control, Athens, Greece, December 1986.

[**BSS80**] M. BETTAYEB, M. G. SAFANOV, AND L. M. SILVERMAN, *Optimal approximation of continuous time systems*, Proc. IEEE, Albuquerque, 1980, pp. 195–198.

[**BB91**] S. BOYD AND C. P. BARRAT, *Linear compensator design: Limits of performance*, Prentice Hall, Englewood Cliffs, NJ, 1991.

[**BEFB94**] S. BOYD, L. EL GHOUI, E. FERON, AND V. BALAKRISHNAN, *Linear matrix inequalities in systems and control theory*, SIAM Publications, Philadelphia, 1994.

[**Con78**] J. B. CONWAY, *Functions of one complex variable*, Springer-Verlag, New York, 1978.

[**CP84**] B. C. CHANG AND B. PEARSON JR., *Optimal disturbance reduction in linear multivariable systems*, IEEE Trans. Automat. Control, AC-29 (1984), pp. 880–887. Tech report dated Oct. 82.

[**C44**] R. V. CHURCHILL, *Modern operational mathematics in engineering*, McGraw-Hill, New York, 1944.

[**CS94**] M. COTLAR AND C. SADOWSKI, *Nehari and Nevanlinna Pick problems and holomorphic extensions in the polydisk in terms of restricted BMO*, J. Funct. Anal., 124 (1994), pp. 205–210.

[**DC80**] C. DEBOOR AND R. CONTE, *Elementary Numerical Analysis: an algorithmic approach*, McGraw-Hill, New York, 1980.

[**DG82**] C. A. DESOER AND C. L. GUSTAFSON, *Design of Multivariable Feedback System with Simple Unstable Plant*, Berkeley ERL Memorandum M82/60).

[**Dor90**] P. DORATO AND R. K. YEDAVALLI, (eds.), *Recent advances in robust control*, IEEE Press, New York, 1990.

REFERENCES

[Do81] R.C. DORF, *Modern control systems*, Addison-Wesley, Reading, MA, 1981.

[Doug72] R.G. DOUGLAS, *Banach algebra techniques in operator theory*, Academic Press, New York, 1972.

[D83] J. C. DOYLE, *Synthesis of robust controllers and filters*, in Proc. of 22nd IEEE Conference on Decision and Control, San Antonio, Texas 1983.

[D$_{report}$] J. C. DOYLE, *Lecture notes in advances in multivariable control*, Honeywell/ONR workshop, Minneapolis, 1984.

[DFT92] J. DOYLE, B.A. FRANCIS, AND A. TANNENBAUM, *Feedback control theory*, Macmillan, New York, 1992.

[DGKF89] J. C. DOYLE, K. GLOVER, P. P. KHARGONEKAR, AND B.A. FRANCIS, *State-space solutions to standard H^2 and H^∞ control problems*, IEEE Trans. Automat. Control, 34 (1989), pp. 831–847.

[DS81] J. C. DOYLE AND G. STEIN, *Multivariable feedback design: Concepts for a classical modern synthesis*, IEEE Trans. Automat. Control, **AC-26** (1981), pp. 4–16.

[Dy89] H. DYM, *J contractive matrix functions, reproducing kernel Hilbert spaces and interpolation*, CBMS Regional Conf. Ser. in Math., 71 (1989).

[Fo91] C. FOIAS, B. FRANCIS, H. KWAKERNAAK, AND B. PEARSON, *Commutant lifting techniques for computing optimal H^∞ controllers*, in: H^∞-*control Theory*, Lecture Notes in Math., 1496 (1991).

[F91] C. FOIAS, B. FRANCIS, J. W. HELTON, H. KWAKERNAAK, AND J.B. PEARSON, H^∞-*control theory*, Lecture Notes in Math., 1496 (1991).

[FF91] C. FOIAS AND S. FRAZHO, *The commutant lifting approach to interpolation problems*, Birkhäuser-Verlag, Basel, 1991.

[F87] B. FRANCIS, *First course in H^∞-control*, Springer-Verlag, New York, 1987.

[FHZ84] B. A. Francis, J. W., Helton, and G. Zames, H^∞-optimal feedback controller for linear multivariable systems, IEEE Trans. Automat. Control, AC-29 (1984), pp. 888–900. Tech. report dated Sept. 1982.

[FOT96] C. Foias, H. Ozbay and A. Tannenbaum, Robust Control of Infinite Dimensional Systems, Springer-Verlag, London, 1996.

[FPE86] G. F. Franklin, J. D. Powell and A. Emami-Naeini, Feedback control of dynamic systems, Addison-Wesley, Reading, MA, 1986.

[G81] J. Garnett, Bounded analytic functions, Academic Press, New York, 1981.

[GMW84] P. Gill, W. Murray, and M. Wright, Practical optimization, Academic Press, London, 1984.

[Gl84] K. Glover, All optimal Hankel-norm approximations of linear multivariable systems and their L_∞ error bounds, Int. J. Control, 39 (1984), pp. 1115–1193.

[GM90] K. Glover and D. C. McFarlane, Robust controller design using normalized coprime factor plant descriptions, Lecture Notes in Control Inform. Sci., 138 (1990).

[Goh64] I. M. Gohberg, A factorization problem in normed rings, functions of isometric and symmetric operators, and singular integral equations, Russian Math. Surveys, 19 (1964), pp. 63–114.

[GL95] M. Green and D. J. N. Limebeer, Linear robust control, Prentice Hall, Englewood Cliffs, NJ, 1995.

[GKL89] G. Gu, P. Khargonekar, and B. Lee, Approximation of infinite-dimensional systems, IEEE Trans. Automat. Control, 34 (1989), no. 6.

[HO94] J.-P.A. Haeverly and M. L. Overton, Optimizing eigenvalues of symmetric definite pencils, Proc. American Control Conference, Baltimore, July 1994.

[H76] J. W. Helton, Operator theory and broadband matching, announced in Proc. of Eleventh Annual Allerton Conference on Circuits and Systems Theory, 1976.

REFERENCES

[H78] J. W. HELTON, *A mathematical view of broadband matching*, IEEE International Conference on Circuits and Systems Theory, New York, 1978.

[H81] J. W. HELTON, *Broadbanding gain equalization directly from data*, IEEE Trans. Circuits and Systems, CAS-28 (1981), no. 12, pp. 1125–1137.

[H82] J. W. HELTON, *Non-Euclidean functional analysis and electronics*, Bull. Amer. Math. Soc., 7 (1982), pp. 1–64.

[H83] J. W. HELTON, *An H-infinity approach to control*, IEEE Conference on Decision Control, San Antonio, Texas, December 1983.

[H85] J. W. HELTON, *Worst case analysis in the frequency domain: The H-infinity approach to control*, IEEE Trans. Automat. Control, AC-30 (1985), no. 12, pp. 1154–1170.

[H86] J. W. HELTON, *Optimization over spaces of analytic functions and the Corona problem*, J. Operator Theory, 15 (1986), no. 2, pp. 359–375.

[H87] J. W. HELTON, *Operator theory, analytic functions, matrices and electrical engineering*, CBMS Regional Conf. Ser. in Math., 68 (1987).

[H89] J. W. HELTON, *Optimal frequency domain design vs. an area of several complex variables: Plenary address*, Mathematical Theory of Networks and Systems, 1989.

[HH86] J. W. HELTON AND R. HOWE, *A bang-bang theorem for optimization over spaces of analytic functions*, J. Approx. Theory, 47 (1986), no. 2, pp. 101–121.

[HMar90] J. W. HELTON AND D. MARSHALL, *Frequency domain design and analytic selections*, Indiana Univ. Math. J., 39 (1990), no. 1, pp. 157–184.

[HMer90] J. W. HELTON AND O. MERINO, *Numerical results in H^∞ control*, in Proc. American Control Conference, San Diego, California, 1990.

[HMer91] J. W. HELTON AND O. MERINO, *Optimal analytic disks: Several complex variables and complex geometry, Part 2*, in Proc. Sympos. Pure Math., Santa Cruz, California, 1991, pp. 251–262.

[HMer93a] J. W. HELTON AND O. MERINO, *Conditions for optimality over $\mathcal{H}^\infty$*, SIAM J. Control Optim., 31 (1993), no. 6.

[HMer93b] J. W. HELTON AND O. MERINO, *A novel approach to accelerating Newton's method for sup-norm optimization arising in H^∞-control*, J. Optim. Theory App., 1993, pp. 553–578.

[HMer98] J. W. HELTON AND O. MERINO, *Classical control using H^∞ methods: theory, optimization, and design*, SIAM, Philadelphia, 1998.

[HMW93] J. W. HELTON, O. MERINO, AND T. E. WALKER, *Algorithms for optimizing over analytic functions*, Indiana Univ. Math. J., 42 (1993), no.3.

[HMW95] J. W. HELTON, O. MERINO, AND T. E. WALKER, *H^∞ optimization and semidefinite programming*, Proc. Conference on Decision and Control, New Orleans, Louisiana, December 1995.

[HMW$_{prep}$] J. W. HELTON, O. MERINO, AND T. E. WALKER, *Semidefinite programming and H^∞ optimization*, J. Robust Nonlinear Control, submitted.

[HS85] J. W. HELTON AND D. SCHWARTZ, *A primer on the H^∞ disk method in frequency domain design control*, Documentation for Fortran software, developed at Lab. for Math. and Statistics, University of California, San Diego, 1985.

[HV97] J. W. HELTON AND A. VITYAEV, *Analytic functions optimizing competing contraints*, SIAM J. Math. Anal., 28 (1997), pp. 749–767.

[Hof62] K. HOFFMANN, *Banach spaces of analytic functions*, Prentice Hall, Englewood Cliffs, NJ, 1962.

[Ho63] I.M. HOROWITZ, *Synthesis of feedback systems*, Academic Press, New York, 1963.

[Hui87] S. HUI, *Qualitative properties of solutions to H^∞-optimization problems*, J. Funct. Anal., 75 (1987), pp. 323, 348.

[JNP47] H.M. JAMES, N. B. NICHOLS, AND R. P. PHILLIPS, *Theory of Servomechanisms*, Radiation Lab. Series, vol. 25, McGraw-Hill, New York, 1947.

REFERENCES

[deWVK78] P. DEWILDE, A. VIEIRA, AND T. KAILATH, *On a generalized Szegö-Levinson realization algorithm for optimal linear predictors based on a network synthesis approach,* IEEE Trans. Circuit Theory, Special Issue on Math. Foundations of Systems Theory, 25 (1978), pp. 663–675.

[Kim84] H. KIMURA, *Robust stabilizability for a class of transfer functions,* IEEE Trans. Automat. Control, AC-29 (1984), pp. 788–793.

[Kim97] H. KIMURA, *Chain scattering approach to H^∞-control,* Birkhäuser, Boston, 1997.

[K83] H. KWAKERNAAK, *Robustness optimization of linear feedback Systems,* IEEE Conf. on Decision Control, San Antonio, Texas, December 1983.

[K86] H. KWAKERNAAK, *A polynomial approach to minimax frequency domain optimization of multivariable systems,* Int. J. Control, 44 (1986), pp. 117–156.

[La89] B. LARSON, *Siso Robust Controller Design via the H^∞ Method,* master thesis, under the direction of Prof. F. Bailey, University of Minnesota, April 1989.

[Le86] L. LEMPERT, *Complex geometry in convex domains,* in Proc. International Congress of Mathematicians, 1986.

[LP61] W. R. LEPAGE, *Complex variables and the Laplace transform for engineers,* McGraw-Hill, New York, 1961.

[L$_{prep}$] K. LENZ, *Properties of Certain Optimal Weighted Mixed Sensitivity Designs,* manuscript.

[LO96] A. S. LEWIS AND M. L. OVERTON, *Eigenvalue optimization,* Acta Numerica, 5 (1996), pp. 149–190.

[M88] O. MERINO, *Optimization Over Spaces of Analytic Functions,* Thesis, University of California, San Diego, 1988.

[M$_{report}$] O. MERINO, *Optimizing real valued functionals on H^1,* manuscript.

[NF70] B. SZ.-NAGY AND C. FOIAS, *Harmonic analysis of operators on Hilbert space,* North Holland, Amsterdam, 1970.

[NN94]   Y. E. NESTEROV AND A. S. NEMIROVSKII, *Interior point polynomial methods in convex programming*, SIAM Publications, Philadelphia, 1994.

[O90]   K. OGATA, *Modern control engineering*, 2nd ed., Prentice Hall, Englewood Cliffs, NJ, 1990.

[OZ93]   J. G. OWEN AND G. ZAMES, *Duality theory for MIMO robust disturbance rejection*, IEEE Trans. Automat. Control, 38 (1993), no. 5.

[RR71]   M. ROSENBLUM AND J. ROVNYAK, *The factorization problem for nonnegative operator valued functions*, Bull. Amer. Math. Soc., 77 (1971), pp. 287–318.

[RR85]   M. ROSENBLUM AND J. ROVNYAK, *Hardy classes and operator theory*, Oxford University Press, 1985.

[S67]   D. SARASON, *Generalized interpolation in H^∞*, Trans. Amer. Math. Soc., 127 (1967), pp. 179–203.

[SS68]   R. SAUCEDO AND E. E. SCHERING, *Introductions to continuous and digital control systems*, MacMillan, New York, 1968.

[SS90]   Z. SHAFIEI AND A. T. SHENTON, *Theory and Application of H^∞ Disk Method*, Report no. MES/ATS/BAE/002/90, Department of Mechanical Engineering, University of Liverpool, U.K., October 1990.

[SI95]   R. E. SKELTON AND T. IWASAKI, *Increased roles of linear algebra in control education*, IEEE Control Systems Magazine, 1995, pp. 76–90.

[Sl89]   Z. SLODKOWSKI, *Polynomial hulls in C^2 and quasicircles*, Ann. Scuola Norm. Sup. Pisa Cl. Sci., 16 (1989), pp. 367–391.

[Sl90]   Z. SLODKOWSKI, *Polynomial hulls with convex fibers and complex geodesics*, J. Funct. Anal., 94 (1990), pp. 156–355.

[SIG98]   R. E. Skelton, T. Iwasaki, and K. Grigoriadis, *A unified algebraic approach to linear control design*, Taylor and Francis, London, 1998.

[SK91]   A. SAYED AND T. KAILATH, *Fast algorithms for generalized displacement structures*, in Proc. Mathematical Theory of Networks and Systems, June 1991, Mita Press, Kobe, pp. 27–32.

[T80]   A. TANNENBAUM, *Feedback stabilization of plants with uncertainty in the gain factor,* Int. J. Control, 32 (1980), pp. 1–16.

[Tr86] L. N. TREFETHEN, *Matlab Programs for CF Approximation,* Numerical Analysis Report 86-3, Dept. of Mathematics, MIT, June 1986.

[VB96] L. VANDEBERGHE AND S. BOYD, *Semidefinite programming,* SIAM Rev., 38 (1996), pp. 49–95.

[V85]   M. VIDYASAGAR, *Control systems synthesis: A factorization approach,* MIT Press, Cambridge, MA, 1985.

[Vit$_{prep}$]   A. VITYAEV, *Uniqueness of solutions of an H^∞ optimization problem in complex geometric convexity,* J. Geom. Anal., to appear.

[We92] E. WEGERT, *Nonlinear boundary value problems for holomorphic functions and singular integral equations,* Academie Verlag, Berlin, 1992.

[Wi41] D. V. WIDDER, *The Laplace transform,* Princeton University Press, Princeton, NJ, 1941.

[Wr97] S. WRIGHT, *Primal-dual interior-point methods,* SIAM Publications, Philadelphia, 1997.

[YJB76a]   D.C. YOULA, H. A. JABR, AND J.J. BONGIORNO, *Modern Wiener-Hopf design of optimal controllers — Part I: The single input – single output case,* IEEE Trans. Automat. Control, AC-21 (1976), pp. 319–338.

[YJB76b]   D.C. YOULA, H. A. JABR, AND J.J. BONGIORNO, *Modern Wiener-Hopf design of optimal controllers — Part II: The multivariable case,* IEEE Trans. Automat. Control, AC-21 (1976), pp. 3–13.

[YS67] D.C. YOULA AND M. SAITO, *Interpolation with positive real functions,* J. Franklin Inst., 284 (1967), pp. 77–108.

[Yng88] N. YOUNG, *An Introduction to Hilbert space,* Cambridge University Press, New York, 1988.

[Z79]   G. ZAMES, *Optimal sensitivity and feedback: Weighted seminorms, approximate inverses, and plant invariant schemes,* Proc. Allerton Conf., 1979.

[ZF81]   G. ZAMES AND B. FRANCIS, *Feedback and minimax sensitivity,* Advanced Group for Aerospace Research and Development, NATO Lecture Notes, no. 117, Multivariable Analysis and Design Techniques.

[ZF83]   G. ZAMES AND B. FRANCIS, *Feedback, minimax sensitivity, and optimal robustness,* IEEE Trans. Automat. Control, AC-28 (1983), pp. 585–601.

[ZDG96]   K. ZHOU, J. DOYLE, AND K. GLOVER, *Robust and optimal control,* Prentice Hall, Englewood Cliffs, NJ, 1996.

Index

bandwidth, 22, 25, 32
    constraint, 23
BodeMagnitude[ ], OPTDesign function, 54, 56
BodePhase[ ], OPTDesign function, 54, 56
bounded function, **3**, 9

Cancel2[ ], OPTDesign function, 52
circular performance function, **38**
closed-loop
    compensator, **5**, 17, 27
    function, **17**
    plant, **5**, 18, 28
    roll-off, 23, 32
        constraint, **24**
    system, **5**, 17, 35, 105, 109
    transfer function, **6**, 101, 106, 109
    optimal, 50
compensator, **5**, 11, 23, 31, 105
    bound constraint, 25, **28**

degree, relative, 3, 104
Design, **7**, 8, 36, 39, 44
designable transfer function, **6**, 35, 43
diagnostics
    Flatness, 58
    Gradient Alignment, 58
    optimality, 58
    output, 58
Discretize[ ], OPTDesign function, 56
disk inequality, **18**, 25, 32, 36, 43

EnvelopeLogPlot[ ], OPTDesign function, 50

EnvelopePlot3D[ ], OPTDesign function, 54, 56
EnvelopePlot[ ], OPTDesign function, 49
external stability, **13**

feedback control, 29
final value theorem, 30, 32
Flat, 51, 58
Flatness diagnostic, 58
frequency band, 18
frequency domain performance requirement, **18**
function
    bounded, **3**, 9
    closed-loop, **17**
    optimal, **38**
    proper, **3**, 9, 35, 106
    rational, **3**, 106
    real, 9, 106
    real rational, 4, 35
    sensitivity, **5**
    stable, **4**, 9
    strictly proper, **3**, 104
    transfer
        closed-loop, 101, 106, 109
        open loop, 18
        open-loop, **5**, 23

gain margin, 19
gain-phase margin, 19, 25, 32, 39
    constraint, **20**
    weighted, **20**
γ_*, 50
gamma*, 51
good performance, **44**
Gradient Alignment diagnostic, 58

INDEX

GrAlign, 51, 58
Grid[ ], OPTDesign function, 56, 59

H^∞ engineering, 115

INT, **101**, 102–104
INT_0, **103**
INT_h, **107**, **109**
INT_h^0, **108**
internal stability, 6, **13**, 15, 31, 35, 105, 106, 109
interpolant, **101**, 102
interpolation condition, 31, **101**, 106, 109
iteration, 58

Laplace transform, 4, 21

MIMO, 115–117
MIMO system, 105

ned, 51, 58
Newton's representation, **102**, 103
Nichols[ ], OPTDesign function, 56
Nyquist plot, 20
Nyquist[ ], OPTDesign function, 56

open-loop transfer function, **5**, 18, 23
OPT, **45**
OPTDesign, 40, 45
OPTDesign[ ], computer output, 50
OPTDParametrize[ ], OPTDesign function, 56
$OPT_\mathcal{I}$, **44**
optimal closed-loop transfer function, 50
optimal function, 38
optimal performance, 38
optimality diagnostic, 58
optimization, **38**
output diagnostic, 58

peak magnitude, 27
    constraint, **27**
performance function, 7, 35–37, **38**, 40, 43

circular, **38**
performance index, **7**, 44
performance requirement, **6**
    bandwidth constraint, 23
    closed-loop roll-off constraint, **24**
    frequency domain, 18
    gain-phase margin constraint, **20**
    tracking error constraint, **22**
phase
    margin, 19
Plancherel theorem, 21n
plant, **5**, 11, 23, 31, 36, 39, 105
    bound constraint, 25, **29**, 32
PlotZP[ ], OPTDesign function, 52
pole-zero cancellation, 9, 11, 14, 35, 105
proper function, **3**, 9, 35, 106

rational fit, 60
rational function, **3**, 106
RationalModel[ ], OPTDesign function, 52
real function, 9, 106
real rational function, **4**, 35
relative degree, **3**, 104
$\mathcal{RH}^\infty$, **4**, 13, 21, 102
RHP-stability, 15
RHPListPlot3D[ ], OPTDesign function, 56
roll-off
    compensator, 58
    rate, 39

sensitivity function, 5, 17
SetGrid[ ], OPTDesign function, 48n
SISO, 115–117
stability
    external, **13**
    internal, 6, **13**, 15, 31, 35, 105, 106, 109
stable function, **4**, 9
strictly proper function, **3**, 104
supremum, **3**

tracking, 32
tracking error, **5**, 17, 21, 25, 30

constraint, **22**, 30, **31**
transfer function, 11, 17
    closed-loop, **6**, 101, 106, 109
        optimal, 50
    designable, **6**, 35, 43
    open-loop, **5**, 18, 23
type n plant, **31**

weight function., 43
weighted gain-phase margin, **20**
Which[ ], Mathematica function, 48
worst-case performance, **38**

yaw rate, 29